For Gabi,

Happy experiments,

Love, Daryl

2019

The Psychophysical Ear

Transformations: Studies in the History of Science and Technology

Jed Z. Buchwald, general editor

Dolores L. Augustine, *Red Prometheus: Engineering and Dictatorship in East Germany, 1945–1990*

Lawrence Badash, *A Nuclear Winter's Tale: Science and Politics in the 1980s*

Mordechai Feingold, editor, *Jesuit Science and the Republic of Letters*

Larrie D. Ferreiro, *Ships and Science: The Birth of Naval Architecture in the Scientific Revolution, 1600–1800*

Kostas Gavroglu and Ana Isabel da Silva Araújo Simões, *Neither Physics nor Chemistry: A History of Quantum Chemistry*

Sander Gliboff, *H. G. Bronn, Ernst Haeckel, and the Origins of German Darwinism: A Study in Translation and Transformation*

Niccolò Guicciardini, *Isaac Newton on Mathematical Certainty and Method*

Kristine Harper, *Weather by the Numbers: The Genesis of Modern Meteorology*

Sungook Hong, *Wireless: From Marconi's Black-Box to the Audion*

Jeff Horn, *The Path Not Taken: French Industrialization in the Age of Revolution, 1750–1830*

Alexandra Hui, *The Psychophysical Ear: Musical Experiments, Experimental Sounds, 1840–1910*

Myles W. Jackson, *Harmonious Triads: Physicists, Musicians, and Instrument Makers in Nineteenth-Century Germany*

Myles W. Jackson, *Spectrum of Belief: Joseph von Fraunhofer and the Craft of Precision Optics*

Paul R. Josephson, *Lenin's Laureate: Zhores Alferov's Life in Communist Science*

Mi Gyung Kim, *Affinity, That Elusive Dream: A Genealogy of the Chemical Revolution*

Ursula Klein and Wolfgang Lefèvre, *Materials in Eighteenth-Century Science: A Historical Ontology*

John Krige, *American Hegemony and the Postwar Reconstruction of Science in Europe*

Janis Langins, *Conserving the Enlightenment: French Military Engineering from Vauban to the Revolution*

Wolfgang Lefèvre, editor, *Picturing Machines 1400–1700*

Staffan Müller-Wille and Hans-Jörg Rheinberger, editors, *Heredity Produced: At the Crossroads of Biology, Politics, and Culture, 1500–1870*

William R. Newman and Anthony Grafton, editors, *Secrets of Nature: Astrology and Alchemy in Early Modern Europe*

Gianna Pomata and Nancy G. Siraisi, editors, *Historia: Empiricism and Erudition in Early Modern Europe*

Alan J. Rocke, *Nationalizing Science: Adolphe Wurtz and the Battle for French Chemistry*

George Saliba, *Islamic Science and the Making of the European Renaissance*

Suman Seth, *Crafting the Quantum: Arnold Sommerfeld and the Practice of Theory, 1890–1926*

Leslie Tomory, *Progressive Enlightenment: The Origins of the Gaslight Industry 1780–1820*

Nicolás Wey Gómez, *The Tropics of Empire: Why Columbus Sailed South to the Indies*

The Psychophysical Ear: Musical Experiments, Experimental Sounds, 1840–1910

Alexandra Hui

The MIT Press
Cambridge, Massachusetts
London, England

MIT Press books may be purchased at special quantity discounts for business or sales promotional use. For information, please email special_sales@mitpress.mit.edu or write to Special Sales Department, The MIT Press, 55 Hayward Street, Cambridge, MA 02142.

This book was set in Stone Sans and Stone Serif by Toppan Best-set Premedia Limited, Hong Kong. Printed and bound in the United States of America.

Library of Congress Cataloging-in-Publication Data

Hui, Alexandra, 1980–
The psychophysical ear : musical experiments, experimental sounds, 1840–1910 / Alexandra Hui.
p. cm. — (Transformations : studies in the history of science and technology)
Includes bibliographical references and index.
ISBN 978-0-262-01838-8 (hardcover : alk. paper) 1. Psychoacoustics—History—19th century. 2. Psychoacoustics—History—20th century. 3. Sound—Experiments—HIstory—19th century. 4. Sound—Experiments—History—20th century. 5. Avant-garde (Music)—History—History—19th century. 6. Avant-garde (Music)—History—History—20th century. I. Title.
QP461.H85 2013
612.8'54—dc23
2012015463

10 9 8 7 6 5 4 3 2 1

For my family

Contents

Acknowledgments

The beneficial conversations ranging from lengthy discussions in graduate school seminars to brief exchanges in elevators at academic conferences are too numerous to individually acknowledge every person that has contributed to this work. Several individuals nevertheless require special acknowledgment. Earlier versions of this book were read by Joel Braslow, Alfred Kramer, Mitchell Morris, Ted Porter, and Norton Wise, truly *mein Doktorvater*. I would also like to thank Horst-Peter Braun, Alison Greene, Paul Josephson, James Kennaway, Julia Kursell, Matthew Lavine, Theresa Leavitt, and Thomas Sturm for valuable criticisms and reactions to portions of this work. The editors at the MIT Press have also provided excellent attention and care, and I must express additional thanks to Elizabeth Judd, my manuscript editor, for her hawkeyed scrutiny. I especially wish to thank Myles Jackson and Peter Pesic for their painstakingly thorough and thoughtful reading of the book manuscript. This book is far better for all of these scholars' advice. I would also like to express my gratitude to Lorraine Daston, for hosting me in Department II at the Max-Plank-Instistut für Wissenschaftgeschichte in Berlin as I completed my archival research as a graduate student. It was the rich and rigorous environment of the MPIWG that sealed my commitment to becoming an academic. Not least, I must thank my colleagues in the Department of History at Mississippi State University as well as the members of the "Long Eighteenth Century" reading group at That School Up North. I am truly grateful for their support and collegiality. Mississippi has become, to my great surprise and delight, an intellectual home.

In Germany, Austria, and the United States I received assistance from the archival staff at a number of libraries and research facilities. Special thanks to the librarians of the Max-Planck-Institut für Wissenschaftgeschichte,

Berlin, Carsten Schmidt of the Bibliothek des Staatlichen Instituts für Musikforschung, Berlin, Jutta Weber and Jean-Christophe Prümm of the Staatsbibliothek zu Berlin, Magrit Prussat of the Deutsches Museum Archives, Steffan Hoffmann at the Universitätsbibliothek Leipzig, the staff of the Österreichische Nationalbibliothek Musiksammlung, Johannes Seidl at the Archiv der Universität Wien, Hermann Böhm and Marcel Atze at the Wienbibliothek im Rathaus, and Kirsten van der Veen at the Dibner Library of the History of Science and Technology of the Smithsonian Institution Libraries. I am also grateful to have received financial support from several sources: a Deutsches Akademisches Austauschdienst (DAAD) Research Grant, a Dibner Library Resident Scholar Award, and a Mississippi State University Humanities and Arts Research Program (HARP) Grant, as well as smaller grants from the UCLA Department of History and the UCLA Neuroscience History Archives.

Some passages in this work have previously appeared in print in the *Annals of Science* and the *Journal for the History of the Behavioral Sciences*, and I am grateful to the original publishers for kindly granting me permission to draw on earlier material of mine.

Finally, it is my pleasure to acknowledge personal debts. The support and friendship of my fellow graduate students at UCLA were central to the development of this work in its earlier stages. I must express special gratitude for the friendship and intellectual inspiration of Naamah Akavia, whose death I continue to mourn. I would also like to thank other friends and especially my family. They have provided love and support as well as thoughtful discussions of my work, and for that I will always be grateful.

Introduction

On a chilly winter evening in 1871, the young professor of experimental physics Ernst Mach delivered a public lecture on symmetry to the rapt audience of the *Deutsches Kasino* in Prague. He explained that repetition of sensations was a source of pleasant feeling. The pleasing effect of visual symmetry was due to this repetition of sensations reflected across an axis. Though this was in part related to the specific structure of the eye, the sense of symmetry must also be "deeply rooted in other parts of the organism by ages of practice," because, for example, an individual's taste in art does not change with the loss of an eye.[1]

But what of the other sense organs? Mach questioned if there was anything similar to the pleasing effect of visual symmetry in the world of sound. Certainly there were repetitions of harmonies and melodies in music but these did not necessarily come from a symmetry of sound. Mach then led his audience through an imagined sound symmetry experiment. He described playing a series of melodies and chord progressions on a piano while first looking in a mirror next to the piano, then placing the mirror below the sheet music and playing the mirrored notes, and finally rotating the sheet music upside down and reading the music from right to left (in Western musical notation, music is read from left to right). He assumed his audience agreed that each of these experiments rendered the melodies unrecognizable and the harmonies transposed from major into minor keys. No symmetry existed, Mach concluded, because a reversal of tones did not result in a repetition of sensations.

You can try it for yourself. Mach has conveniently written it out in a manner that doesn't require a mirror (figure 0.1). Take the first passage, for example. Play the top line from left to right. To reverse the passage, play it from right to left. You might notice that the rhythm remains symmetrical,

Figure 0.1
"Musical cut" for Mach's sound symmetry experiments. Ernst Mach, "On Symmetry," in Ernst Mach, *Popular Scientific Lectures*, 3rd ed., trans. Thomas McCormack (Chicago: Open Court, 1898), 102.

but music is not rhythm alone. Further, even in this case of symmetrical rhythm, the symmetry is only apparent over time, not in an instant. To invert the passage harmonically, play the top line again from left to right and then the line below it from left to right. To both reverse and invert the passage, play the top line again from left to right. Then, rotate the entire page upside down and play what is now the bottom line, from left to right. If you've had some musical training you might further notice the transposition of the harmonies from major to minor. This transposition from major to minor was, Mach explained, like inversion for the eye. It was symmetry in the mind, not the sensations. The unintelligible sounds heard during this awkward exercise on the piano illuminated something about the structures and processes of the mind for Mach.

I happened upon this lecture by Mach when researching his acceleration sense experiments for a seminar paper in graduate school. I admired this charming demonstration and immediately sat down at my Yamaha upright to play it myself. While struggling to read music from right to left, I conjured the image of Mach sitting down at his own piano, a Bösendorfer grand, sportingly plunking his way through the music passages, just to be sure.

Pondering the lecture again, I found it curious that Mach's discussion could so easily move from aesthetics to sound sensation to music theory and back. Further, I was impressed that his audience was savvy enough to keep up. This was a world in which the sensation of sound and the experience of music were interchangeable. A psychophysical experiment could also be an aesthetic one. I asked how this could be. I asked what that audience had heard. This book is an attempt at an answer.

In the mid-nineteenth century, with the construction of new concert halls and the creation of new compositional trends, the concertgoers of the great cities of Europe were exposed to new rhythms and harmonies. Not only were the sounds they heard novel, but their hearing itself was also new and different. From 1840 to 1910, leading physicists, physiologists, and psychologists were preoccupied with understanding the sensory perception of sound from a psychophysical perspective. They sought the direct and measurable relationship between physical stimulation and psychical sensation. This book examines this little-known, yet formative moment in the middle of the nineteenth century when the worlds of natural science and music coalesced around the psychophysics of sound sensation. New musical aesthetics were intertwined with new conceptions of sound and new conceptions of hearing.

This entanglement of the aesthetics of music and the psychophysics of sound sensation can be located in the psychophysicists' conscious and unconscious efforts to incorporate specific sounds into their experiments—that is, musical sounds. And not just any musical sounds but the musical sounds listened to by upper-middle-class, liberal Germans and Austrians. The psychophysicists approached their project with the songs and strains of the music culture of the late nineteenth-century German-speaking world ringing in their ears. As a consequence, the psychophysicists, in their efforts to understand the sensory perception of sound, struggled with theoretical issues that also plagued the world of music theory and musical aesthetics such as form, temperament, the primacy of the Western tonal system, and the possibility of aesthetic evolution.

At issue for the psychophysicists was the seeming irreconcilability of universal physical and physiological laws with the historical and cultural contingency of musical aesthetics. Did different peoples in different times and places create different music because they had different ears? Because they heard differently? Such possibilities were at odds with the idealist, universalizing goals of their science. This tension is at the core of my twofold argument. I claim that not only was the nineteenth-century psychophysical study of sensory perception framed in terms of musical aesthetics, but it became *increasingly* so. As musicology and musical aesthetics shifted to prioritize and celebrate the individual listener's experience, the psychophysicists collapsed experimental subject with experimental object, accepting that subjective, individual experience was the only avenue through which to study sound sensation. Sharing a common language of musical aesthetics, psychophysicists and musicologists also shared a common turn toward focusing on the aural experience of the individual.

From the 1840s through the first decade of the twentieth century, the exploration of the sensory perception of sound was intimately related to a series of shifts that occurred in both science and music. The study of the sound sensation changed from the earliest forms of psychophysical study—of sensory perceptual threshold or just-noticeable-difference experiments—to focus on the processes of hearing. Then, in another shift, the psychophysicists studying the sensory perception of sound became increasingly interested in the completely subjective and individual experience of the listener/experimental subject. This shift can be seen as increasingly intertwined with aesthetics, first in a very general and cultural way, but by the end of period examined, both in music circles and psychophysical circles, the study and understanding of sound sensation was framed in terms of the individual experience of the experimental subject. Rather than becoming increasingly *wissenschaftlich* and narrowly preoccupied with experiment, psychophysicists were very much engaging the intersection of sensory perception, music culture, and subjectivity. Instead of being increasingly disengaged, which is generally the thrust of the natural sciences at the end of the nineteenth century, the scientific study of sound sensation actually became more bound up with music culture over time.

Cultural shifts outside the immediate sites of the psychophysical studies of sound sensation also informed the practices and goals of the psychophysical investigators, resulting in a changing cultural niche. In nineteenth-century music circles two aesthetic camps dominated: the first emphasized slow, rich tones and harmonies, compositions and performers that showcased such sounds, and music critics that buttressed this aesthetic with historical, political, and moral arguments in their favor. The second camp celebrated virtuosic performance and compositions that presented impossibly fast passages or provocative harmonies. They too had critics that supported their efforts both philosophically and politically. These groups turned to the sciences for support and even, at times, presented their own *wissenschaftliche* experiments and arguments. At stake was the definition of musical sound and musicianship.

This struggle was further exacerbated by the transition to equal temperament from older tuning systems and the introduction of non-Western and *Volk* music, which together suggested a possible plurality of valid tonal and aesthetic systems as well as the convergent development of musical aesthetics. Tones and harmonies that had previously been accepted as fixed began to shift and change. The new sounds were both exciting and jarring. And they threatened to undermine the conception of sound sensation as universal as well as Western musical aesthetics as the most fully evolved. The composers of the early twentieth century pushed this destabilization of the Western tonal system even further, actively dismantling its harmonies and instead emphasizing the subjective experience of the individual. Ultimately, this was a renegotiation of the relationship between critic, composer, performer, and listeners. It was also an acknowledgment of the individual, subjective listening experience.

Loosely defined as the study of the relation between physical stimulation and psychical sensation, psychophysics was a subdiscipline shared by several different disciplines (physics, physiology, and psychology) or perhaps an "orientation."[2] It was also brand new, based on an epiphany experienced by Gustav Fechner in 1850, presented a decade later in his publication of *Elemente der Psychophysik* (*Elements of Psychophysics*), and breathlessly embraced by its practitioners for both its experimental and philosophical (a reunification of the mind and body!) possibilities. This book focuses on the psychophysical work of Gustav Fechner, Hermann Helmholtz, Ernst Mach, Wilhelm Wundt, and Carl Stumpf. Some of these individuals would have considered themselves psychophysicists. Some would have considered only their experimental approach psychophysical and would have maintained for themselves, for example, the title of physicist. To be sure, these individuals shared a firm belief that the theories and techniques of several disciplines needed to be combined in order to fruitfully study sensory perception.

Further, the psychophysical practitioners not only drew on multiple disciplines in their work, but incorporated values and meanings and *sounds* from beyond the walls of the laboratory. Current histories of psychophysics, most within the history of psychology, reduce psychophysics to a story of the origins of the psychophysical law, also known as the Fechner-Weber law. This is, in part, a product of the current status of psychophysics as a subfield of experimental psychology. Presently, psychophysics is little more than the experimental application of the Fechner-Weber law. But from the middle of the nineteenth century through the beginning of the twentieth, it was much more.

The psychophysical investigators this book examines were trained in a variety of disciplines and yet their experimental studies of sound sensation shared many common elements. First, all assumed sound and music to be equivalent. This assumption did not necessarily include all music, but specifically Western music. Until the end of the nineteenth century, when colonial expansion exposed the European public to non-Western culture, Western music was the only music anyone had ever heard. In their studies of sound sensation the psychophysical investigators employed Western musical instruments as scientific instruments and, correspondingly, Western musical tones and intervals as sound. In this way, again, the psychophysical study of sound was always also a psychophysical study of musical aesthetics.

Further, the psychophysical investigators were all competent if not highly skilled musicians. They read music and played instruments. They had to. In the nineteenth century if people wanted to hear music, they had to either attend a concert or perform it themselves. Music had to be performed to be experienced. In this way, psychophysicists themselves were instruments that both generated and evaluated the sound experience. This was both a requirement of, and only possible at, the intersection of psychophysics and music. No other nineteenth-century art medium or experimental projects embodied and mobilized subjective experiences this way.

I highlight these features of psychophysical experiments—that sound was Western music, that musicianship was a requirement of experimental practice—to indicate the extent to which the music world and the natural scientific world overlapped in this period. It was more than an intellectual exchange, but a social and a cultural exchange as well. Every one of the psychophysicists I examine circulated with ease in the music world. They were friends with professional musicians, music critics, musical society directors, and composers. The psychophysicists' use of music in their research, as both an experimental object and an experimental technique, lacked defensive, explanatory discussion. This was, I argue, indicative of the extent to which music and musical aesthetics were integrated into their understanding of the natural world. To employ music in their studies of sound sensation was a straightforward and unselfconscious impulse. It was a function of their way of hearing. The psychophysicists drew on the resources of their material culture. Their conceptions of hearing were a product of what and how they heard the world around them.

There were three major tensions, common to both the work of the psychophysicists and the musicologists, at play in this narrative. First, the meaning of expertise, both scientific and musical, was in flux. Because psychophysics was not a traditional and well-defined discipline and therefore not necessarily bound to a set of disciplinary practices—indeed it was initially conceived as a unification of many practices—its boundaries were fluid. Psychophysicists had the freedom to traverse boundaries both within the sciences and outside, in culture more broadly. And because the individual, subjective experience of music was not only valid but necessary for the psychophysical studies of sound sensation, all or at least *more* practitioners could be experts. For a time, for some natural scientists, the natural scientific work of musicologists, music critics, even musicians was accepted as valid. Musical skill and the musician's ear were considered scientific skills. Just as psychophysicists stepped into the world of music for experimental subjects and experimental instruments, so too did individuals from the music world explore experimental studies of sound sensation.

For others, the lack of firm disciplinary boundaries was threatening and unstable. For them, expertise was a means of staking out intellectual territory and enforcing borders. Fierce debates followed over what skills were required to competently psychophysically study sound sensation—essentially, over what kind of ear should be studied and used for study. In tension, then, was the issue of what kind of expertise was relevant for psychophysical studies of sound sensation: musical, experimental, or some combination of both.

All the psychophysical investigators of sound sensation described in this book also struggled to reconcile universal laws of sound sensation with the variable laws of musical aesthetics. This second tension was inherent in the framing of the experimental approach. As discussed earlier, the psychophysical investigators treated sound and Western music as equivalent. To varying degrees they all accepted that Western musical aesthetics were historically and culturally contingent, that they were specific to time and place, likely the product of an evolutionary dynamic. So their studies sought not only a conception of sound sensation that was universal but an explanation for why musical aesthetics were contingent. Psychophysics and the subjective experience of music could potentially reconcile the universal and contingent, but they also contributed to the instability of the very foundations of the sound sensation project.

The third tension was similarly rooted in anxiety about the scientific goal of searching for universal laws. Within the natural sciences generally, substantial restructuring occurred at the end of the nineteenth century, in Germany especially. Disciplines and institutions divided and proliferated, fueling intellectual and material turf wars and identity crises. Within experimental science an additional shift was taking place, away from experimental design informed by a search for universal laws (so the experimental subject was interchangeable and part of the instrumentation) toward a preoccupation with the interpretive or analytical skills of the experimental subject. This was a shift in emphasis toward the subjective individual rather than the universal. For a time, many psychophysicists studying the sensory perception of sound attempted to reconcile universality with subjectivity in their work. Their efforts took different forms, informed by what they believed to be at stake, by their personal tastes, their own musicianship, their friendships, and their disciplinary training. We can understand this shift in experimental design, indeed in the very goals of natural science, as a reevaluation of the role and skills of the scientist.

Increasing awareness and study of the individual, subjective experience of listening fueled a growing historicism among psychophysicists studying sound sensation (I therefore encourage the reader to attempt the described sound sensation demonstrations and experiments whenever possible—your individual, subjective experience of sound was precisely what the psychophysicists were interested in). This paralleled developments in both physics and the life sciences. The irreversibility of thermodynamic systems, evolutionary biology, and ecology: all suggested a contingency of phenomena. There were perhaps universal laws of development for aggregate systems but not of causality on an individual level. For those studying sound sensation, there was a growing belief that hearing itself was historical. For some, knowledge itself was bound to time and place, a possibility that had significant consequences for the practice of science.

In locating and contextualizing the psychophysical study of sound sensation, this book follows in the tradition of works that historicize sensory perceptual phenomena that come into being as both valuable techniques of experimental investigation and epistemic objects of inquiry, specifically Jonathan Crary (historicizing perception), Michael Hagner (historicizing attention), and the joint effort of Lorraine Daston and Peter Galison (historicizing scientific objectivity).[3] By showing convergence of aesthetics and psychophysics, rather than divergence, my book is part of a recent trend in the history of science that challenges older portrayals of the fin-de-siècle as a period of incommensurability and disintegration. Here, I refer to Deborah Coen's and Suman Seth's respective works.[4] Both of these texts find the instability of the intellectual world during the fin-de-siècle to be rich and fertile rather than a period of crippling crisis. I find similar trends in my analysis of the late nineteenth-century psychophysical impulse of the German-speaking world.

Many fine historical studies have addressed specific aspects of the German psychophysical impulse, but they consider the role of music in only an incidental way, if at all.[5] There have been, however, some very good histories that examine the relationship between studies of sensory perception and the visual arts.[6] These works lay the foundation for a growing body of work in the history of science that examines how image making is knowledge making.[7] The emphasis of these works on the centrality of the practice and material culture affiliated with the visual arts in natural science is, in my work, extended to the music arts. I examine the ways the natural sciences drew on the practice and material culture of the music world, so, I examine how music making is knowledge making.

My book is thus part of an effort within the history of science that builds on the cultural history works of the previous decades focused on the sociopolitical elements of musical reception in the long nineteenth century.[8] These studies in the history of science explore points of intersection between this expanded realm of music and science and technology. Emily Thompson's wonderful book, *The Soundscape of Modernity: Architectural Acoustics and the Culture of Listening in America, 1900–1933*, presents a rich examination of sound spaces.[9] Elfrieda Hiebert and Erwin Hiebert, as well as David Pantalony, detail the intellectual innovations that developed from Hermann Helmholtz's interaction with musicians and musical instrument makers.[10] Myles Jackson's *Harmonious Triads: Physicists, Musicians, and Instrument Makers in Nineteenth-Century Germany*, pushes beyond detailing the sites of interaction to the very practice of intellectual exchange between natural scientists and musical instrument manufacturers.[11] These works, in expanding the role and reach of music to the material culture of science and technology, reinforce what musicologists have long claimed: that music was the dominant means of artistic expression in the nineteenth century. Given the primacy of music as art form as well as the extent to which psychophysical investigators interacted with musicians and musicologists, it is remarkable that the interaction between psychophysical studies of sound sensation and musical aesthetics has never been the central focus of historical analysis. This book provides that analysis.

The first history-of-science monograph devoted to the life story of psychophysics, *The Psychophysical Ear: Musical Experiments, Experimental Sounds, 1840–1910* shows how the practice of psychophysics was from its very beginnings directed toward answering aesthetic questions. The psychophysical study of sound sensation was also a study of musical aesthetics. This book is therefore a cultural history of science. The intended audience is students and scholars in the modern history of science, modern cultural European history more generally, and musicology. To many, some of the individual works or debates or issues discussed will be familiar. That is to be expected. One of the larger goals of this book is to unite several well-known narratives, to highlight not just points of intersection but reinforcement, and to show how these seemingly disparate intellectual and cultural developments can be better understood when framed in relation to psychophysical studies of sound sensation.

I have referred to this book as the "life story" of psychophysics, or at least the sound sensation project of psychophysics, but this is not to say that the narrative ends with a death. Instead the respective projects of the psychophysics of sound sensation and musicology converged. Each chapter of the book examines an episode in which psychophysics and music intersected, brought together by common assumptions and tensions, bound together by material culture and understood through a series of exemplary vignettes in the life story of psychophysics. These vignettes, presented in overlapping chronology, also develop spatially, expanding from a single psychophysicist to an ever-expanding network of natural scientists, philosophers, composers, musicians, musicologists, and field ethnomusicologists.

Chapter 1 begins with Gustav Fechner and Ernst Heinrich Weber's development of the psychophysical law, Fechner's aesthetic philosophy, and his friendship with the leader of the Leipzig music scene, Hermann Härtel of Breitkopf und Härtel music publishers and piano manufacturers. Other members of the Leipzig music scene as well as its distinctive performance culture are also introduced. These narrative strands of Fechner's life are interwoven to show how, given Fechner's extensive interests in aesthetics as well as his connections to the Leipzig music scene, the project of psychophysics was an aesthetic one from the very start.

Chapter 2 is an examination of three centrally important figures in musical aesthetics, music criticism, and musicology in the second half of the nineteenth century: A. B. Marx, Eduard Hanslick, and Hugo Riemann. This chapter explores the relationship between their respective aesthetic theories of form in musical composition (here mostly Marx and Hanslick) and understanding of the individual listener. I also focus on Riemann's undertone experiments in order to illustrate the efforts of individuals in the music world to supplement and legitimize their work in music criticism and musicology with natural scientific "evidence." Further, I show that this foray of musicologists into natural science, at least in the case of sound sensation study, was acceptable, if only briefly.

In the third chapter, I show how the project of psychophysical aesthetics of sound could be literally embodied in a single individual: Hermann Helmholtz. Helmholtz's own musical training, his classicist tastes, and his familiarity with the manufacture of musical instruments allowed him to reconcile his goals of describing sound sensation in terms of universal physical and physiological laws and his awareness that musical aesthetics were variable across cultures and time. Helmholtz's musical experience is contextualized within larger debates about aesthetics as well as musical performance and reception between the formalist musicians associated with Leipzig and the New German School cultivated by Franz Liszt and Richard Wagner.

In the fourth chapter, I trace the close collaboration between Ernst Mach and the Viennese music critic Eduard Kulke on the sensory perceptual phenomenon of accommodation in hearing. For both Mach and Kulke, the accommodation in hearing had implications for an understanding of musical aesthetics—specifically, Kulke's love of Wagner—which was developmental and followed evolutionary laws. And for Mach, this led to his belief, by the early 1860s, that hearing itself was historical.

The progression of these vignettes reveals that as both the music world and the natural scientific world changed, the psychophysical investigators' efforts to reconcile universal laws of sound sensation with musical aesthetics required increasingly historicist understandings of both musical aesthetics and sound sensation. So it is fitting that the last chapter indicates how psychophysics began to evolve into an entirely different entity. I use the debate between Wilhelm Wundt and Carl Stumpf over their respective tone differentiation experiments to show that, in the realm of physiological psychology, the goals of psychophysical studies of sound sensation were also moving away from issues of musical aesthetics. This disappearance of the aesthetic dimension of psychophysics resulted both from the professionalization of such disciplines as experimental psychology, musicology, and ethnomusicology, and from the proliferation of new and different musical aesthetics at the beginning of the twentieth century.

1 Gustav Fechner, the Day View, and the Origins of Psychophysics

$g = k\log(\mathrm{b}/b)$

—Gustav Fechner, *Elemente der Psychophysik*[1]

Every one of the individuals involved in psychophysical studies of sound sensation from the middle of the nineteenth century on into the twentieth, responded to the accomplishments of Ernst Heinrich Weber and Gustav Fechner. Their work, Fechner's especially, provided an empirical framework for psychophysics as a potential means of collecting quantitative data on the processes of sensory perception. Carl Stumpf described Fechner and Weber as "great men, genuine scientific investigators" who had made lasting impressions on him.[2] He saw his own work on tone differentiation at the end of the nineteenth century as a defense of the applicability of the Fechner-Weber law to sound sensation. Ernst Mach and Hermann Helmholtz certainly both referred to portions of their experimental work as psychophysical. Helmholtz and Fechner communicated briefly. Wilhelm Wundt was a colleague at the University of Leipzig as well as a great admirer. On the hundredth anniversary of Fechner's birth, Wundt gave a lengthy speech on his psychophysical worldview. Wundt described Fechner as "the reformer (*Erneuerer*) and perfector (*Vollender*) of nineteenth-century romantic natural science."[3]

William James described Fechner as the master of Wundt's entire generation of experimental scientists and credited Fechner with establishing the foundation of both scientific psychology and experimental aesthetics.[4] In his introduction to the English translation of Fechner's *Das Büchlein von Leben nach dem Tode* (*The Little Book of Life After Death*), James described Fechner's mind as "one of those multitudinously organized cross-roads of truth, which are occupied only at rare intervals by children of men, and from which nothing is either too far or too near to be seen in due perspective."[5] James had his own motivation for embracing Fechner—as a foil to late nineteenth-century English and American transcendental philosophy —but his suggestion that Fechner's enduring relevance lay in his ability to generate cross-disciplinary conversations was apt. To varying degrees, Mach, Helmholtz, Stumpf, and Wundt engaged the cosmological components of Fechner's psychophysical monism. And they all struggled with the aesthetic implications the psychophysical approach to sound sensation unleashed.

When the Fechner-Weber law is employed in experimental psychology today it is done so devoid of its romantic trappings. This decontextualized understanding of the Fechner-Weber law means that the contemporary understanding of psychophysics itself is reduced to the law only. And discussions of the history of psychophysics are discussions of the history of the Fechner-Weber law only.[6] But to psychophysicists from the middle to the end of the nineteenth century, the law held far-reaching potential. The practice of psychophysics in the nineteenth century was from its very beginnings oriented toward addressing aesthetic questions. The Fechner-Weber law traditionally used to mark the origins of psychophysics—and reductively and mistakenly used to *define* psychophysics—followed from Fechner's monistic, "day-view" science and was bound up with his experimental aesthetics.

An examination of both Fechner's early writings advocating a *Tagesansicht* or "day-view" approach to science and Weber's early experimental work on touch sensitivity, reveals that Fechner's psychophysical worldview was well established prior to his development of the law itself. Fechner drew on Weber's strictly physiological study of subjective perception to different ends. For Fechner, subjectivity could be exploited experimentally to understand differences in aesthetic judgment. Ultimately, Fechner saw his psychophysics as a tool for an experimental aesthetics. It was this aesthetic potential—as well as a central, indeed critical role for subjective observation—that gave psychophysics the appeal and flexibility to be viably employed as an experimental practice for over half a century.

The Leipzig Music World

Fechner arrived in Leipzig in 1817 to attend the university and never left. At the time it was still a town; the population was contained entirely within the old walls. While Fechner was by no means a central figure in the Leipzig arts scene, his diary entries indicate that he was close to many individuals who were. His brother and nephew were both artists and both his niece and her stepdaughter were concert pianists. His wife, née Clara Volkmann, is credited with introducing him to polite Leipzig society. Fechner occasionally attended concerts, sometimes a small gathering at a friend's home, sometimes the opera. Though Fechner repeatedly claimed in his diary entries to lack expertise and even understanding of music, he had carefully read both Eduard Hanslick's *Vom Musikalisch-Schönen* and Hermann Helmholtz's *Die Lehre von den Tonempfindungen als physiologische Grundlage für die Theorie der Musik* (see chapters 2 and 3 for an extended discussion of these texts).[7] Additionally, despite his humility about his musical expertise, Fechner always had an opinion about the numerous musical performances he attended. Most of these discussions were negative—that the pieces or types of singing were not to his taste or that he found operas lacking arias of the Mozart or French melodic style very boring.[8]

Fechner admitted that while he found music interesting, he had no musical expertise.[9] Again, the larger goal of this chapter is to show how Fechner's psychophysical program was also an aesthetic one. In his psychophysical work, he shied away from examining music and sound sensation specifically. Direct connections between a psychophysics of sound sensation and music in Fechner's work are few. It remains valuable, however, to map out the musical world that surrounded Fechner before proceeding to a discussion of his psychophysical and aesthetic projects.

If his diaries are any indication, Fechner's exposure to music appears to have been guided by social interactions rather than an independent interest. He had two personal connections to individuals central in the music world. The first was his sister's daughter's stepdaughter, the virtuoso pianist Clara Wieck. In 1831, barely eleven years old, she gave her first concert at the Leipzig Gewandhaus. The following year she performed in Paris, launching her international career as a celebrated performer. In 1840 she married the composer Robert Schumann to become Clara Schumann.

Robert Schumann and friends had, in 1834, founded what they termed the *Davidsbündler* (David Guild) to lead the reformatory fight against musical Philistines. Their mouthpiece was the journal *Neue Zeitschrift für Musik*, founded in the same year and edited by Schumann until 1844.[10] Schumann's contributions to the journal were often critical of the public affection for flashy and virtuoso performances. In 1835, at Clara's home (he had studied piano with Clara's father and Fechner's niece's husband, Friedrich Wieck), he had been introduced to Felix Mendelssohn, who had just been made conductor of the Leipzig Gewandhaus Orchestra. Both men would work to establish a unified musical canon and revive interest in composers of the past: Mozart, Beethoven, Weber, and Bach. In 1854, suffering from tinnitus and demonic visions, Schumann attempted suicide. Following his rescue (he had jumped from a bridge), he asked to be taken to Richarz's sanitarium in Endenich, where he died in 1856. Clara would continue her career as a concert pianist and maintain control of her husband's musical legacy, both through the performance of his pieces in her concerts and as editor of his works with Breitkopf und Härtel.

In the last decade of his life, Schumann, though sensitive and withdrawn, had been embraced as a leader of the Germanic music world. His death thus seemed all the more tragic and contributed to the lasting image of Robert Schumann as the shy, tortured romantic. An article by the Viennese music critic Eduard Hanslick, "Robert Schumann in Endenich," perpetuated this portrayal, to the point that Friedrich Nietzsche later attacked Schumann as representative of the degenerative effects of romanticism in music.[11] Clara, in a letter to Johannes Brahms, fretted over Hanslick's piece, especially the intimate details of Robert's illness and the personal letters that were included.[12] She was outraged that the music critic would presume to write about the illness, especially since Hanslick had been such an advocate for her and her husband.

Much has been written on Hanslick's alignment with Leipzig, the city of Mendelssohn, Schumann, and Brahms, against Wagner. Hanslick was deeply critical of both the rising Neudeutsche Schule (New German School) that coalesced around Wagner and Liszt in the 1850s as well as the early to mid-nineteenth-century cult of the virtuoso that contributed to this ascension. Musicologist James Garratt describes a complementary phenomenon in the cult of the composer.[13] Garratt contrasts the virtuoso, symbolic of bad individualism, with the composer, who was "elevated as a figurehead for the artistic and socio-political aspirations of the participants," and presents Felix Mendelssohn as emblematic.[14]

Mendelssohn, already known regionally as a musical prodigy, had gained worldwide acclaim for his revival performance of the all-but-forgotten *St. Matthew Passion* by J. S. Bach with the Berlin Singakademie in 1829.[15]

Scholars consider this performance to mark the beginning of the repertoire movement, a series of revival concerts of "great works" by past German composers. On becoming conductor of the Leipzig Gewandhaus Orchestra, which claims to be the oldest continuing symphony orchestra in the world, Mendelssohn was critical to the city's development into a world-class musical center. He used the orchestra as well as the opera and the choir of Thomaskirche (where J. S. Bach had been choir director) to showcase both historical works and those of contemporaries, his own included. It was at the commemorative music festivals that, Garratt argues, the adulation of Mendelssohn as citizen-king genius was most apparent. The festivals allowed for the reconciliation of the concept of artistic genius as autonomous individual with communal action and the projection of middle-class aspirations of social mobility and meritocracy.[16] A creature of the *Vormärz*, the festival composer-conductors symbolized a new kind of leader. Their authority was granted legitimacy and yet constrained by the festival participants.[17]

Fechner's other personal connection to the music world was his close friend Hermann Härtel, of the firm Breitkopf und Härtel that specialized in manufacturing pianos as well as in publishing books and sheet music, which it continues to this day.[18] The firm was also responsible for the publication of the *Allgemeine musikalische Zeitung*, the leading German-language music journal of the nineteenth century, from 1798 to 1848 and again from 1866 to 1869. Härtel's position as the head of the dominant music publishing company at the time meant that he was in constant correspondence with every significant composer, critic, and performer in Europe. His correspondents included Moritz Hauptmann, Clara and Robert Schumann, Franz Liszt, Johannes Brahms, and Josef Joachim. The content and extent of the Härtel's correspondence with Joachim—Joachim was updating him on his travels and concert programs rather than simply soliciting publication—suggest that the two were quite close (more on both Joachim and Clara Schumann in chapter 3). Not too surprisingly considering that the Breitkopf und Härtel firm also built pianos, Härtel was in touch as well with several acousticians, including Ernst Flourens Friedrich Chladni, Ernst Heinrich, and Wilhelm Weber (more on all of these individuals shortly).[19]

Härtel was also a member of the Leipziger Kunstverein and hosted large music parties at his home in which both local and foreign "musical notables" were frequently in attendance. In his diaries, Fechner described attending an 1844 gathering hosted by Härtel that included the composer Felix Mendelssohn, Clara and Robert Schumann, as well as the violinists Ferdinand David, Heinrich Wilhelm Ernst, and Joseph Joachim, among others.[20] The city of Leipzig, in part because it hosted both the Schumanns and the house of Härtel, was an international music center and Fechner was in the middle of it.

Measuring Sound

The city of Leipzig was also a significant site for the scientific study of sound. The story of sound as an investigative object necessarily begins in Leipzig for it was there, at the end of the eighteenth century, that sound was made measurable. Ernst Florens Friedrich Chladni is presently best known for his technique for rendering sound visible, his "Chladni figures," presented in his 1787 book, *Entdeckungen über die Theorie des Klanges* (Discoveries about the Theory of Sound). Chladni sprinkled a metal plate with metal filings or sand and then, with his fingers placed at various locations on the plate, bowed it perpendicularly. Depending on where he placed his fingers and where he bowed the plate, the filings or sand distributed themselves into a variety of lacy symmetrical figures (figure 1.1). Chladni's contemporaries celebrated his experiments on the acoustics of vibrating plates as groundbreaking.[21] The beautiful sand figures that appeared on the surface of the vibrating plates could be measured, the correlation between vibration frequency and nodal points quantified with precision. Chladni had made sound visible—and, as a consequence, made sound more readily experimental.

Chladni's subsequent experiments on the properties of elastic curved surfaces as well as sections of his second major work, *Die Akustik*, drew on his ability to quite literally see sound in a new way.[22] Myles Jackson describes Chladni as "the nodal point between acoustician and musical-instrument maker."[23] His position at the intersection of the worlds of music and science—literally mirroring the nodal points of his Chladni figures, Jackson points out—combined with his new techniques for examining sound, established the framework through which nineteenth-century studies of sound sensation were pursued: experimental and bound up with aesthetics.

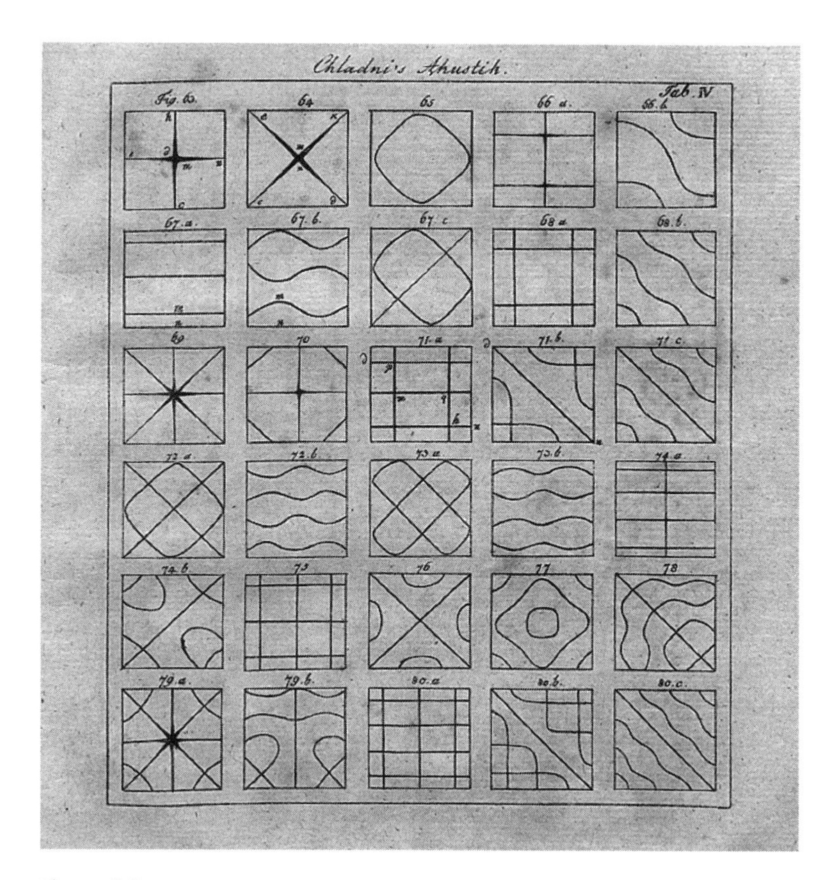

Figure 1.1
Chladni figures. Engraving from Chladni's *Die Akustik* (1802).

At the beginning of the nineteenth century, Chladni was a lodger and frequent guest with the Weber family in Wittenberg. Two of the twelve Weber children, Wilhelm and Ernst Heinrich, can also be considered nodal points between the worlds of music and the worlds of science. Like Chladni, Wilhelm Weber, originally trained in mathematics, worked at the intersection of acoustics and musical instrument design.[24] Wilhelm and Ernst Heinrich together in 1825, with Chladni's encouragement, published *Die Wellenlehre auf Experimente gegründet*, a presentation of their experimental work on the properties of standing light and sound waves. The brothers dedicated it to Chladni. Chladni, in turn, published a gushing summary review of the book in the music journal *Cäcilia* the following year.[25]

The Weber brothers continued their experimental work over the next decade but their interests diverged. Wilhelm continued to examine the acoustics of organ reed pipes. Ernst Heinrich, trained as a physician and holding a chair in anatomy at Leipzig, followed more physiological pursuits, namely the physiology of the eardrum and touch sensation. Ernst Heinrich Weber's subsequent study of sense discrimination—just-noticeable differences of touch, temperature, weight, and later vision, smell, and pitch—was resonant of Chladni's acoustics. Ernst Heinrich Weber made sensory perception experimental.

Ernst Heinrich Weber's 1834 book, *De pulsu, resportione, auditu et tactu.* The final section, entitled *De subtilitate tactus* (usually referred to as *De Tactu*) presented a series of experimental investigations of the sense of touch and his conclusions about the cause of variation in touch sensitivity.[26] The "touch organ" sensed, Weber began, both pressure or traction and temperature. His experiments examined the variation over the body in sensitivity to difference of location of points of stimulation, difference of weight, and difference of temperature. For example, in measuring the touch organ's ability to distinguish weight difference between hands, Weber had the experimental subject place his or her hands on a table.[27] A piece of cardboard was then placed as a screen between the subject and his or her hands (and Weber's work). Equal weights were placed on both of the subject's hands. Then, without the subject's knowledge, both weights were lifted and one replaced with a heavier one before both were placed back down on the subject's hands. The experimenter then lifted the weights off both hands and either switched the weights between the right and left hand or placed them back down without switching them. Weber noted if the subject correctly distinguished the heavier weight from the lighter. In this way he could quantify the sensitivity to weight difference of each experimental subject. The experimental subjects in this particular experiment included not only Weber himself but also "a merchant unskilled at weighing objects," "an educated man, a mathematician," "a girl," "a woman," and "a student of literature."[28] Other experiments were performed exclusively on and by Weber and his brother, Wilhelm.[29]

From his extensive experimentation on the touch organ, Weber concluded that both deliberate movement of the body and physiological structure of the touch organ contributed to tactile acuity. He elaborated that it was the structure of the touch organ that was responsible for variation in sensitivity to pressure or temperature. Weber explained that areas of greater sensitivity were also areas with higher density of nerves. Further, variation distributed along longitudinal or transverse axes could be due to the direction in which the nerve was oriented.[30] Throughout *De Tactu* Weber sought to root his discussion of the touch sense in the physiological structure of the touch organ.

Weber's experimental approach to physiology was a move away from that of the other leader in his field, Johannes Müller at the University of Berlin (Müller would later be Helmholtz's *Doktorvater*). Müller had argued that sensation was experienced only through the effect of "specific nerve energies" on the sensory nervous system. Although he distinguished between different "modalities" of experience, Müller maintained that the activity of the sensory nerve (due to the effect of the specific nerve energies) underlay all sensory experience. Discarding much of this language of specific nerve energies, Weber sought to explain the touch sense in terms of physiological structure alone.

In the following decade Weber further developed this line of thinking as he expanded his research program to include sensory perception in general. In *De Tactu* he had distinguished between objective and subjective perceptions. Objective perception arose when an awareness of both the stimulation of one of the sense organs and the object generating the stimulation occurred. Weber gave the example of distinguishing between light and the source of the light.[31] A subjective perception was a perception of a stimulated sense organ only and not of the stimulating object separately. These subjective perceptions could only arise when the stimulating object came in direct contact with the body.[32] These two forms of perception are thus distinguished through the interpretation of the sensation: an objective perception resulted in an awareness of an object separate from the body; a subjective perception resulted in an awareness of the body's own sensory states, usually pain or pleasure. This subjective perception, an awareness of the body's own sensory state, was termed *Gemeingefühl* (common sensibility). Weber devoted his 1846 essay, "Der Tastsinn und Gemeingefühl" (On the Sense of Touch and Common Sensibility), to demonstrating that this Gemeingefühl was rooted in the physiological structure of the central nervous system.

Weber's experiments on individuals' sensitivity to differences of location, weight, and temperature on the skin led him to make two claims about the Gemeingefühl. First, individuals were better able to compare successive than simultaneous sensations.[33] Weber believed that finer discrimination was prevented in the cases of simultaneous stimulation by the merging of the two impressions. This phenomenon of the summation of sensations obscuring discrimination was most pronounced in his studies on heat sensitivity. Weber argued that this phenomenon indicated that all comparisons of sensory impressions were between a past and present impression. Perceptions were not recalled from memory but instead sustained even after the cause of the impression was gone. When a new impression was made it obscured the perception of the earlier one so that nothing of it remained except its discrepancy from the second.[34] This best explained the greater sensitivity to successive rather than simultaneous impressions. Recall that much of Weber's work was on the touch organ, where he poked at the skin with a stylus. So this language of sustained impression was quite literal as well as highly physical and psychological.

Weber also concluded that individuals, when distinguishing between sensations, did not perceive the absolute difference between the impressions, but instead the ratio of the difference of their magnitude.[35] In *De Tactu* he had shown that for distinguishing between tones, sensitivity varied by ratios of vibration frequency, not absolute differences between tones.[36] In "Der Tastsinn" Weber revisited and reiterated this psychological phenomenon of comparing relative differences rather than absolute differences between sensations. He cited additional studies to argue that the phenomenon occurred across the senses, in vision and weight perception as well as hearing. Then, suggesting that this perception of ratios rather than absolute differences led to a physiological basis for aesthetics, Weber asserted that "in music we can perceive tonal relationships without knowing the number of vibrations involved: in architecture we can conceive of spatial relationships without having to measure them in inches."[37]

Weber maintained his belief that perceptible impressions were rooted in the physiological structure of the central nervous system, specifically the structures of the nerves and the brain. In fact, it was only the areas of the body equipped with nerves that had Gemeingefühl.[38] Weber cited studies in which the severing of nerves of a specific area corresponded to a loss of Gemeingefühl in the same area. Also, Gemeingefühl was the most acute in areas with the greatest density of nerves. Weber's work on touch sensitivity in *De Tactu* had led him, a decade later, to better understand subjective perception. For Weber this subjective perception was very narrowly defined as Gemeingefühl, an individual's awareness of his or her own sensory state.

"Day-View" Science and Psychophysical Monism

Following his matriculation at the University of Leipzig to study medicine in 1817, the young Gustav Fechner diligently attended Ernst Heinrich Weber's physiology lectures. The two men eventually became close friends. In 1834, during the period in which Weber was refining his experimental study of sensory perception, Fechner became a full professor of physics at the university. Throughout the 1820s and 1830s, in addition to his textbook writing and experimental work on electricity, Fechner published a series of pseudo-satirical pieces under the pseudonym "Dr. Mises." These writings, with such titles as *Beweis, daß der Mond aus Jodine bestehe* (Proof That the Moon Is Made of Iodine), "Vergleichende Anatomie der Engel" (The Comparative Anatomy of Angels), and *Das Büchlein vom Leben nach dem Tode* (The Little Book of Life After Death), were intended as satirical criticism of the medical profession. The latter text was Dr. Mises' only serious work and presented, among a number of almost mystical beliefs, Fechner's conviction that individual consciousness continued beyond death. Fechner, as Dr. Mises, devoted "Ueber einige Bilder der zweiten Leipziger Kunstausstellung" (On some Paintings at the Second Leipzig Exhibition) to critiquing the wrong path of idealization. He later claimed that this piece, from 1839, marked the beginning of his interest in an experimental aesthetics.[39] Which is all to say that by the end of the 1830s, Fechner was struggling mightily to reconcile his physics research with his interest in phenomenalism, in the possibility of a psychical world beyond the physical one, and in aesthetics generally.

In 1839 Fechner gave up his chair in physics as he succumbed to a lengthy and bizarre illness in which he became nearly blind and emaciated to the point that he could no longer stand. Light pained his eyes and he could barely eat. Fechner also complained of a mental fatigue and what he eventually came to see as a battle between his self and his thoughts.

A few years later, Fechner described his illness to his nephew Johannes Kuntze as follows:

My inner self split up as it were into two parts, my self and my thoughts. Both with each other; my thoughts sought to conquer my self and go an independent way, destroying my self's freedom and wellbeing, and my self used all the power at its will trying to command my thoughts, and as soon as a thought attempted to settle and develop, my self tried to exile it and drag in another remote thought. Thus I was mentally occupied, not with thinking, but with banishing and bridling thoughts. I sometimes felt like a rider on a wild horse that has taken off with him, trying to tame it, or like a prince who has lost the support of his people and who tries slowly to gather strength and aid in order to regain his kingdom.[40]

Fechner's recovery marked a reconciliation of sorts, the unification of an intellectual life that had previously been split, and, as the passage above suggests, in conflict during his illness. By 1846 he was no longer publishing his more flamboyant tracts under a pseudonym. He resumed lecturing, though on topics that departed from his original physics chair such as the seat of the soul, as well as psychophysics and aesthetics.[41] William James later explained that Fechner believed his faith, both religious and cosmological, had saved him from his illness and that it was this belief that motivated him to forthrightly communicate the nature of his faith to the world.[42]

In 1851 Fechner presented his fully developed conception of panpsychism—his previous works merely hinted at various elements of it. In *Zend-Avesta oder über die Dinge des Himmels und des Jenseits* (Zend-Avesta or on the Things of Heaven and the Beyond) he introduced his concept of the soul (*die Seele*) of the Earth, or all-inclusive consciousness of which all individual human consciousnesses (and those of all animals and plants) were constituent parts.[43] The universe everywhere was alive and humming and conscious. Elaborating on his earlier understanding of immortality (as presented in *Das Büchlein des Lebens nach Tode*), Fechner explained that after death, the perceptive contributions of the individual organism live on as part of the greater consciousness, growing and forming new relations forever.

Further, argued Fechner, the distinction between the physical, material, corporeal world and the *geistige* (loosely translates as not-necessarily-religious spiritual) was only a matter of perspective and linguistic convention. There was a parallelism between the corporeal and geistige, and, correspondingly, a means of observing what appeared to be physical from instead an inner point of view in which subject and object were combined.[44] Fechner critiqued strictly materialist science as reductive and one that limited knowledge of the world to human subjectivity. In terms of immediate facts and experiences, the individual could only know his or her own consciousness. Yet, knowing that one's consciousness was connected to the all-inclusive, immortal consciousness, allowed for a vast analogical series of knowing.[45] Through analogy and the principle of connection (*Zusammenhang*), much could be known about both the material and the geistige.[46]

Fechner later termed the narrow, materialist view of the world *die Nachtansicht* (the night view). *Die Tagesansicht*, the day view, in contrast, was an understanding of nature that moved beyond a strictly materialist approach.[47] Fechner's day-view science maintained a direct realism in that he believed physical appearances existed objectively in the world and were not simply the products of subjective consciousness. These appearances, however, were interconnected through the highest consciousness to the psychical aspect of human nature.[48] To view the world only in terms of its material features was to overlook the connection of all things, both corporeal and geistige, objective and subjective. This understanding of the world—in which physical and psychical experiences were two different perspectives on the same event, two sides of the same reality—anticipated the monism Fechner would articulate a decade later in *Elemente der Psychophysik*.

Philosopher-historian Michael Heidelberger has described Fechner as "a radical empiricist with a phenomenalistic outlook," and his day-view philosophy of nature as a "non-reductive materialism."[49] Fechner's conception of day-view science was an extension of his belief that consciousness was independent of life. Because consciousness extended beyond the human lifespan, it was not limited to the physical world. Some inner experiences might never develop beyond the psychical world (to affect or be affected by the physical world). His theory of psychophysical monism allowed him to reconcile his empiricism with his phenomenalism. In his efforts to establish experimental support for his monism, Fechner looked to the touch sensitivity studies of Ernst Heinrich Weber, his physiology professor and friend. Fechner, however, used Weber's experimental results for different ends: an experimental foundation for his new monistic, day-view science of psychophysics.

The Law

Fechner's day view did not, as might be assumed, preclude rigorous experimental practice. In fact, according to Fechner, those practitioners that combined the psychical and the physical in their approach were to be commended. Good science would exploit the parallelism of the psychical and physical worlds. In an undated letter (though likely 1879) to Hermann Härtel, Fechner included a numbered list of individuals that employed exemplary day views.[50] (See figure 1.2.) The list included "Professor Ludwig," "Professor Wundt," and "Professor Volkmann."

In *Zend-Avesta*, Fechner recounted his musings on how to establish day-view science in a solid empirical form on the morning of October 22, 1850.[51] He had been lying in bed when it struck him that an arithmetic series of psychical "intensities" likely corresponded to a geometrical series of physical "intensities." He then realized that Weber's 1846 study of the touch sensitivity supported just this relationship (equal relative stimulus increments corresponded to equal sensation increments) and presented it in mathematical form. What Fechner called Weber's law—following the 1860

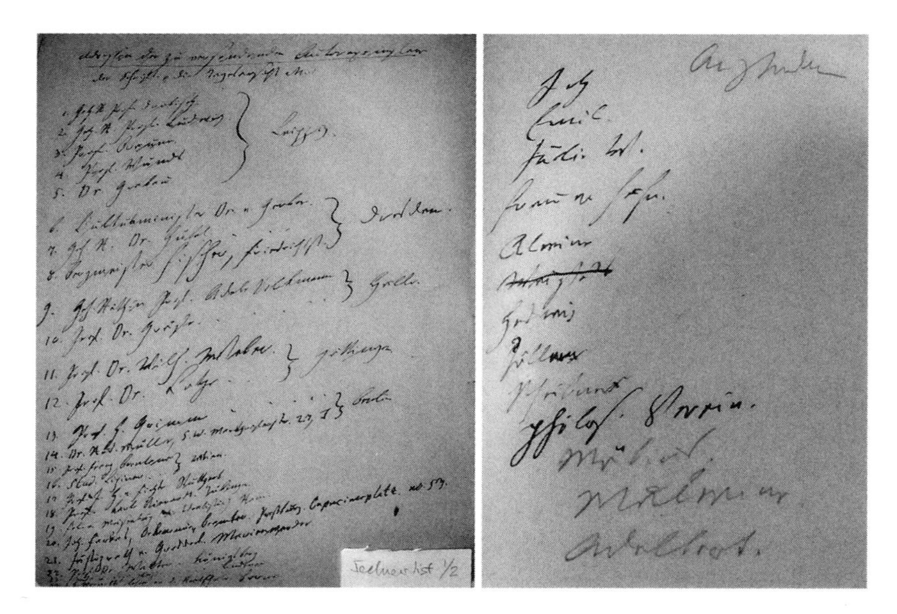

Fechner list 1/2

Figure 1.2
Fechner's list of individuals with exemplary day views

publication of *Elemente der Psychophysik* it would be termed the Fechner-Weber law—was intended to be the empirical basis of his new day-view science of psychophysics. The chronology should be noted: Fechner had already laid out the framework of his monistic, day-view science before he turned to Weber's research as the crucial experimental support. This indicates just how important it was to Fechner to ground his psychophysics in experiment. Additionally, it shows that Fechner's psychophysics was more than just the Fechner-Weber law. It was also the most explicit expression of his monism.

Fechner's description of his October 1850 epiphany followed a discussion in his diary detailing his search for a mathematical expression connecting the body (*Leib*) and soul (*Seele*). He had, he explained, already written volumes about the relationship but his work needed a mathematical foundation.[52] Fechner believed the consequences of his discovery (*Entdeckung*) could be tremendous for all of philosophy and noted that he should remember the date. His project then was to reconcile this mathematical theory with the conclusions of his previous work in *Zend-Avesta*, which had been incomplete.[53] Looking to Weber's work on touch sensitivity was a good place to start and, as both Weber's student and longtime friend, Fechner would have known Weber's work well.

Elemente der Psychophysik was the culmination of Fechner's effort to articulate his monism in mathematical form.[54] Fechner claimed that the implications of his psychophysical work could not possibly support a position on the mind-body problem because the work itself was a consideration of the functional relationship between the two sides.[55] Further, as an exact science like physics, psychophysics "must rest on experience and the mathematical connection of those empirical facts that demand a measure of what is experienced or, when such a measure is not available, a search for it."[56]

In the second volume of *Elemente der Psychophysik*, Fechner presented just such a mathematical connection. He stated that Weber's law, in which equal relative stimulus increments corresponded to equal sensation increments, should be regarded as fundamental to the program of psychophysics. If b were the stimulus, and a small increase were db, the relative increase would be db/b, corresponding to an incremental increase in sensation dg. According to Weber's law dg remained constant when db/b was constant. Further, as long as dg and db/b remained very small, their changes would remain proportional to one another. The relation between the two could therefore be expressed:

$$dg = k\,db/b,$$

where k is a constant dependent on the units of g and b.[57] This mathematical formula did not presuppose or establish the measurement of sensation. It was simply an expression of the relationship between small relative stimulus increments (db/b) and sensation increments (dg). By integrating this expression, Fechner arrived at the logarithmic "functional formula between stimulus and sensation":

$$g = k(\log(b/\mathit{b}))$$

In this formula, *b* denotes the threshold value of the stimulus, at which sensation g begins and disappears. In this form, psychophysics was not simply the general articulation of the functional relation between the physical and the psychical, but an experimental program.[58] Fechner believed that it was possible to exploit this functional relation between the psychical and physical experimentally.[59] Physical stimulation caused psychical sensation via the action of internal bodily processes that, according to Fechner, followed general laws. Therefore one could learn of these lawful internal bodily processes, the lawful relationships between psychical sensation and physical stimulation, through physical measurement.[60] This was the task of outer psychophysics: to identify "the truly basic empirical evidence for the whole of psychophysics . . . in as much as it is only this part that is available to immediate experience."[61] Indeed, much of the first volume of *Elemente der Psychophysik* was devoted to empirically demonstrating this outer psychophysical relation for touch, weight, pitch differentiation, and intensity of light, color, and sound. Much of the experimental support for the fundamental laws and facts of psychophysics, was, in fact, Weber's work. Fechner cited large passages from Weber's publications, including data tables. The experiments that Fechner performed himself built on Weber's experiments. Fechner saw them as follow-up work, a means of developing new experimental methods and perhaps of finding exceptions to Weber's general law.[62]

Psychophysical Aesthetics "von Unten"

In this chapter thus far I have shown that Fechner's psychophysics was a product of his monism. He had already established his monistic, day-view science by 1850 when he realized, lying in bed on the morning of October 22, that Ernst Heinrich Weber's work on touch sensation was just the experimental support he required. Fechner's use, however, of Weber's experimental results was decidedly different from Weber's. Weber had turned to an exploration of the physiological roots of Gemeingefühl, the subjective perception of one's own sensory state. In the decades following his publication of *Elemente der Psychophysik*, Fechner also became increasingly interested in subjective perception, but in its role in aesthetic judgment. The psychophysics of subjective perception was Fechner's entry point to the experimental study of aesthetics. Fechner mobilized his psychophysics as the basis for his experimental aesthetics "from below" (*von Unten*).

Fechner took a more radical position on the role of individual, subjective experience in experimentation on sensory perception than Weber. For Fechner, the subjective, psychical world and the physical world were not simply bound together, they were the same world, two sides of the same reality. Though he did not believe that the psychophysical law could directly contribute to such higher mental processes as aesthetics, Fechner did believe that some aesthetic frameworks could be linked to psychophysically measurable sensations. His monism allowed him to exploit the subjectivity of experimental observers to a greater degree.

In *Elemente der Psychophysik* Fechner made a distinction between what he termed "inner psychophysics" and "outer psychophysics." Inner psychophysics addressed the relation of the psychical realm to the internal functions of the body. Outer psychophysics examined the relation of the psychical realm to phenomena external to the body. The difference between inner and outer psychophysics was rooted in what he saw as a difference between such higher mental activities as thought, will, and finer aesthetic feeling and the lower mental activities of sensation and drive. Although the higher mental activities were "exempt from a specific relationship to physical processes," they remained bound by a general relationship.[63] Though the ability to study inner psychophysics was quite limited then, psychophysics would eventually need to address the relationship between higher mental processes and their physical foundations.

According to Fechner, sound sensation provided one current entry point to examine the relationship between inner and outer psychophysics. Psychophysical studies of sound intensity and pitch discrimination could inform an understanding of harmony and melody in music. The experience of harmony and melody—what Fechner considered higher mental processes—was based on the ratios of vibrations, which the listener perceived as separate sensations. Thus, although harmony and melody were features of the finer aesthetic feelings of higher mental processes, they were still bound to sound sensation and therefore measurable though psychophysics.[64]

Fechner addressed the measurement of both sound intensity (volume) and pitch discrimination, but only briefly. He cited the experiments on the limits of audibility done by Karl Emil von Schafhäutl, another individual that embodied the intersection of music and natural science in this period. Schafhäutl's extensive work in theoretical and practical acoustics was often done in collaboration with the flute virtuoso and wind instrument maker, Theobald Boehm. Though he turned to metallurgical experiments in the 1830s, becoming a professor of geology, mining, and metallurgy at Munich University in 1841, Schafhäutl maintained an interest in music and acoustics. He corresponded about English organ making and church music for the *Allgemeine musikalische Zeitung* while pursuing a patent lawsuit in England, reported on musical instruments at trade exhibitions, and was an official organ examiner for nearly forty years.[65]

Schafhäutl's audibility experiments had determined that the limit of audibility was the sound of a cork ball of 1 mg, falling a distance of 1 mm at midnight, when no wind was blowing. Further experimentation with "young people, whose ear had been musically trained," produced similar results.[66] Schafhäutl thus concluded that the 1 mg cork ball dropped from the height of 1 mm could "be taken as the acoustical energy which marks the average limit of sound intensity just perceivable by the healthy human ear subject to the influences of our civilization."[67]

For pitch discrimination, Fechner cited work by Félix Savart on the limits of determining pitch difference. Savart is perhaps best known for his electrodynamics work with Jean-Baptiste Biot on the magnetic field generated by an electric current, the Biot-Savart law. His great expertise, however, was in the physics of vibrating bodies and he, along with Chladni, experimentally confirmed the theory of vibrating rods developed by Daniel Bernoulli and Leonhard Euler. He also expanded on Chladni's sand figure experiments and developed acoustical instruments including a siren.

The French mathematician and physicist, Charles Delezenne, combined his specialties and applied calculus to acoustics. In 1837 he developed a rotating tone wheel capable of producing a sustained oscillating electrical current, one of the very first precision electroacoustical instruments. Of his studies on the just-noticeable differences of the consonance of tones, Fechner pointed out that Delezenne's experiments actually focused on the experimental subject's ability to note the deviations from pure musical intervals, which did not directly address the psychophysical law. But, Fechner continued, there was no need to demonstrate experimentally that the psychophysical law applied to pitch discrimination. He claimed that it was, after all, the "simple—even the notorious—testimony of a musical ear, that equal ratios of vibrations correspond to equally large differences between tones at different octaves."[68]

Fechner also compared ratios of intensity and pitch, asking "several people with a good musical ear" to compare two sounded pitches of differing intensity and pitch.[69] Remarkably, those willing to try the experiment agreed that the pitch interval was a fourth. But Fechner was hesitant to give the conclusions much validity because the experimental setup had been quite crude and provisional.[70]

Fechner's work did, however, suggest that a psychophysical study of pitch discrimination could provide insights into musical aesthetics. Superior pitch discrimination ability appeared to correlate with "good musical ears." Although aesthetics was one of the higher mental processes confined to the realm of inner psychophysics, Fechner did allow that some aesthetic frameworks could be accessed by outer psychophysics (based on the Fechner-Weber law) through measurable sensations. Harmony and melody, Fechner explained, were based on the ratios of the vibrations that underlay separate tone sensations. These vibration ratios changed only in an exact and dependent relation to the combination and series of sounded tones.[71] Musical aesthetics could be understood through a study of sound sensation.

In the decades following the publication of *Elemente der Psychophysik,* Fechner intensively engaged with aesthetic theory, mostly the visual arts. He published articles on his psychophysical study of the golden section (Carl Stumpf was actually one of Fechner's experimental subjects for this project) as well as a comparison of the two Holbein Madonnas.[72] In the latter study, along with other points of comparison, Fechner sought to establish experimentally which of the Holbein Madonnas—the one exhibited in Dresden or the one exhibited in Darmstadt—was more beautiful.

When the two images were displayed together, Fechner conducted what amounted to an exit poll of visitors' judgments.[73]

Fechner's 1876 book, *Vorschule der Aesthetik* (Introduction to Aesthetics), was the culmination of this line of work, introducing the foundational principles of an experimental aesthetics.[74] The title betrays the publication's programmatic intent. Fechner's efforts to experimentally understand the basis of sensory perception (*sinnliche Wahrnehmung*) and aesthetic judgments (*ästhetische Urteile*) resulted in a call for an aesthetics that established its principles inductively, that was an aesthetics from both above and below.[75] Fechner's guiding question was not what was pleasing but *why* it was pleasing.[76] Beauty itself was his experimental object.

Two features of the six aesthetic principles (*Principe*) at the core of Fechner's psychophysical aesthetics are worth highlighting. First, he called the first principle the "principle of aesthetic threshold." The similarity of this language to that employed by Weber in his touch sensitivity studies is noteworthy. And second, Fechner's sixth and main aesthetic principle, the "aesthetic association principle," was bound up with memory (the association part) and individual, subjective experience.[77] This association principle came into play in the case of music. In the only part of the book devoted to music, Fechner explored the possible reasons certain moods (happiness, sadness, and so on) were associated with certain harmonies and melodies. He was inconclusive, though memory was certainly implicated.[78]

Historian Uta Kösser argues that what distinguished Fechner's aesthetics from similar discussions by his contemporaries was that Fechner did not force a demarcation between *Natur-* and *Kulturwissenschaften*.[79] Taking this a step further, I would argue that for Fechner it was not a case of allowing boundaries between natural science and aesthetics to blur. Rather, for Fechner the two realms were often the same. Psychophysics was the foundation of his aesthetics.

Conclusion

On the surface, Weber and Fechner appear to have been studying sensory perception with different goals in mind. Weber sought—and believed he had found—strictly physiological foundations for the sensory perception processes he studied. Fechner pushed Weber's results farther, using them as empirical support for his day-view science of psychophysics and, ultimately, his psychophysical aesthetics. But they shared a common interest in subjective perception. And they were also, in a sense, unified by the data, Weber's data. Further, they both insisted on experimental support for their claims. As a study of the origins of psychophysics, this chapter has highlighted the fact that the psychophysical program from its very beginnings was experimental and interested in subjective perception.

Moreover, Fechner believed that psychophysics could be used to better understand aesthetics generally. Though aesthetics was part of the realm of higher mental processes and therefore not directly accessible through outer psychophysics, aesthetics was related to such lower mental processes as sensation. Outer psychophysics could evaluate sensations. For Fechner it was a means by which aesthetics could be studied experimentally, even quantified. Psychophysics as framed by Fechner was an aesthetic project from its very origin. This made it a particularly good resource for natural scientists interested in examining aesthetic systems through sensory perception. Indeed, this possibility was likely the source of much of the initial enthusiasm for psychophysics following Fechner's publication of *Elemente der Psychophysik*.

The phenomenalist roots of psychophysics also built in fluid boundaries between experimental subject and experimental object. Direct observation was of the highest value. The experimental subject's sensations could, arguably, be prioritized to the point of abstraction. It is unlikely that Fechner would have allowed this, but the point is worth pursuing. An experimental framework that prioritized the subjective experience of the experimental subject—like psychophysics would—could result in the abstraction of the experimental subject to the point that the psychophysical investigators were also the experimental objects themselves. This flexibility of subject and object, observer and observed, proved extremely fruitful for the psychophysical investigators of sound sensation discussed in the following chapters.

2 From Sonically Moving Forms to Inaudible Undertones: The New Musical Aesthetics of A. B. Marx, Eduard Hanslick, and Hugo Riemann

One is trying to force Nature to sound undertones, which cannot exist according to the laws of mechanics, and the most recent dualist Dr. Hugo Riemann tells us that he and nobody else, save perhaps for Aristotle two thousand years ago, has heard these undertones, which alone are supposed to explain the consonance of the minor triad.

—Karl von Schafhäutl[1]

The German-speaking musical world in the middle of the nineteenth century was highly unstable. New tuning systems, new tones, new music theories, and the fledgling discipline of musicology all jostled to establish position. The transition from earlier forms of tuning to equal temperament meant the pitches themselves were not fixed, neither standardized between instruments nor within individual ones. Later, a growing interest in non-Western music introduced entirely new sounds.

This sonically unstable world put new demands on the listener. The musicologist Lawrence Kramer argues that autonomous music had also created its own aesthetic crisis of sorts.[2] The elimination of the connection between text and voice to the point that music was conspicuously autonomous resulted in a palpable incompleteness.[3] This analysis does little, however, to explain the overwhelming popularity of autonomous music in the nineteenth century. Most listeners found what Kramer terms semantic emptiness to be of great aesthetic value, and thus the act of closely following the autonomous musical form was considered one of the highest types of aesthetic experience. Indeed, much nineteenth-century music criticism was a reexamination of the proper form of listening.

With new interest in the individual listener came the possibility of the subjective, individual experience of sound. The mid-nineteenth-century sonic instability meant that many acoustic and music-theoretical claims were, at least initially, valid—no one individual or group of individuals controlled the creation of knowledge. There was, at least for a short while, the potential for dialog between psychophysical investigators of the natural sciences and members of the music world. This provided an opportunity for the fledgling discipline of musicology to present a new framework through which to understand the changing world of music.

The musicologist Hugo Riemann is an extreme example of the consequences of the open season on sonic experience. He was the only person that heard harmonic undertones, the sonic mirror of the harmonic overtone series. Riemann argued that because *he* could hear the undertones, his theory was valid. Only a single witness was necessary and it was acceptable for the single witness to be Riemann himself. For a brief period, both music theorists and natural scientists accepted Riemann's subjective listening experience as support for his music-theoretical system of harmonic dualism.

This chapter is an examination of three centrally important figures in musical aesthetics, music criticism, and musicology in the second half of the nineteenth century: A. B. Marx of Berlin, Eduard Hanslick of Vienna, and Hugo Riemann of Leipzig. Through an exploration of Marx and Hanslick's respective aesthetic theories of form in musical composition and analysis and their evolving understandings of the role of the individual listener, I show how it was possible for Riemann to claim to hear harmonic undertones. Further, I show how it was possible for others, both in the worlds of music criticism and natural science, to believe him.

If the first chapter set the psychophysical scene in relation to the larger cultural world, then this chapter establishes the musicological scene in relation to the larger natural scientific world. The final section of this chapter addresses the early years of the discipline of musicology in light of Riemann's actions seeking and then rejecting a natural scientific basis for his music-theoretical theories. Musicology, though it initially sought the credibility associated with the natural sciences, instead prioritized formal analysis and historical approach over the acoustic and psychophysical. The implication is that the exchange between the natural scientific and music world was not equal. The psychophysicists took far more from music than the music critics and musicologists took from science.

An Unstable Sonic World

In 1834, Johann Scheibler, the inventor of the tuning fork tonometer, lobbied at a Stuttgart congress of physicists for the standardization of concert A at 440 cycles per second.[4] The development of new acoustical instrumentation, the tuning fork tonometer perhaps the single most important, had brought great promise of standardization and equivalence both within individual instruments and between them. Equal temperament, with its associated freedoms of composition and performance, was practicable, no longer mere theory.[5] And yet, in a paper read before an 1880 Society of Arts meeting in Britain, Alexander Ellis stated: "Equal temperament is now generally aimed at, though seldom really attained."[6] Ellis then listed various well-known pianos and organs that had and had not been tuned to equal temperament (not a single English organ at the 1851 Great Exhibition, for example, was equal tempered). Even by his 1885 English translation of Hermann Helmholtz's opus, *Die Lehre von den Tonempfindungen als physiologische Grundlage für die Theorie der Musik* (On the Sensations of Tone as a Physiological Basis for the Theory of Music), Ellis's notes and appendixes indicate that equal temperament as presently understood was not practiced on pianos. It would not become common practice until the twentieth century.[7]

Earlier Western tuning systems struggled with the problem of a twelve-tone scale in which intervals were based on pure ratios that were repeatable over many octaves.[8] Twelve acoustically pure fifths are almost, but not quite, equivalent to seven acoustically pure octaves. A twelve-tone scale in which all tones were related by pure fifths could not be enclosed within an octave. The solution had been to choose certain intervals—the most commonly played ones—to be pure and to sacrifice others. The Pythagorean tuning system, used through the Renaissance, prioritized fourths and fifths as pure. To do so, the Pythagorean "comma," or the microtonal discrepancy between twelve perfect fifth ratios and seven octaves, had to be distributed among the thirds and sixths so that they were slightly larger than pure. Compared to pure thirds or sixths, these Pythagorean-tuned intervals would have sounded quite jarring and out of tune. A later tuning system, just intonation, instead prioritized triads (and therefore thirds) as well as fourths and fifths as pure, distributing the Pythagorean comma elsewhere. While such tuning systems meant that only certain intervals and therefore certain keys (of diatonic scales) could be performed on the instrument without a complete retuning, those keys that could be performed were said to have distinctive coloration and character of sound, a quality coveted by composers and listeners.[9] Many lamented the impure intervals and the associated loss of key coloration with the tempering of the Western scale. Helmholtz, for one, bemoaned: "I do not know that it was so necessary to sacrifice correctness of intonation to the convenience of musical instruments."[10] He even went so far as to document the beats —his criterion for dissonance—of the tempered triad, implying that equal temperament was the most unnatural of the many tuning and tempering options.

Though equal temperament had promised great freedom for composers and performers alike (concert programs or even single works could modulate between a greater variety of keys without a retuning of the instruments), in practice at the end of the nineteenth century, it was inconsistently applied. This was partly due to the fact that while the means of measuring equal temperament did exist, the techniques for equal tempering a piano or organ of eighty-eight or more keys with precision did not. The art of traditional tuning practices based on aesthetic judgments of each tone by ear could not be applied successfully to the science of equal temperament. Ellis's survey of organs, operas, pianos, and old tuning forks across Europe found the pitch of the concert A—supposedly standardized in response to Scheibler's efforts at the 1834 Stuttgart conference—ranging from 373 cycles per second on Delezenne's French Foot Pipe in Paris to 567 cycles per second on the Praetorius church organ in northern Germany.[11] To our present, equal-tempered ears, centered on a 440 Hz A, a 373 Hz tone would be somewhere between an F$^\#$ and a G and a 567 Hz tone would be between C$^\#$ and D. There was no consistency between the instruments of Europe, in either tuning systems or individual tones.

In addition to shifting tones within the Western tuning system, sounds altogether new to Europe were being introduced. Increasingly, non-Western, "primitive" music ensembles visited Europe to perform.[12] Additionally, the introduction of the phonograph to field studies in the late 1870s granted music the ability to travel; field ethnomusicologists returned to Europe with wax cylinders full of never-before-heard music. This non-Western music, some of which was based on highly complex scale systems, undermined European beliefs in the inherent superiority of their Western intonation. The introduction of the non-Western music also fueled the development of new questions and theories about the origins of Western harmony. This will be examined more fully in chapter 5.

With the Western tuning system in flux, with tones themselves unstable, an individual's claim about what he or she heard (or did not hear) was negotiable. It was certainly possible for an individual to hear something that others did not. This lack of reproducibility, of course, was not tolerated for long. The subjective experience of sound was, only fleetingly, a valid means of making universal claims about sound sensation. It was, however, centrally important to the fledgling field of musicology, which was simultaneously redefining the act of listening to music. To this development I now turn.

A New Analytical System

Adolf Bernhard Marx, Berlin's dominant mid-nineteenth-century music theorist and critic and, for a time, close friend of Felix Mendelssohn, was instrumental in the codification of idealist sensibilities in music theory and performance repertoires. As the head editor from 1824 to 1831 of the new weekly *Berliner allgemeine musikalische Zeitung*, developed to compete with the popular *Allgemeine musikalische Zeitung* of Leipzig, Marx was in a position to define and cultivate a new listening public in the city. Which he actively attempted to do throughout his career.[13] Marx's project reflects the more general mid-nineteenth-century classical belief dominant in the German-speaking lands—though most strongly seen in Prussia—that education, especially in art, was essential for the proper cultural development of both the individual and the nation.

Central to Marx's compositional and pedagogical theories was his concept of form in music. For Marx, form and content were bound together. Content was only intelligible "through the agency of form."[14] It was impossible for form in art to exist independently, for itself only.[15] Musical form was none other than the determination of content, the self-realization of the *Geist*.[16] The implication of this was that form, constantly changing in the process of self-determination, was developmental, even historicist. Marx framed this, interestingly, in terms akin to a biological imperative:

> Indeed, it appears that we must welcome this alteration, this variability of forms. The world hungers for the new or, at least, for change in its pleasures. Genius must stride forward, talent strives after it; that which is inwardly unfinished struggles despairingly in the feeling of its emptiness and impotence for some refashioned or even misshapen form, the "circulation of matter" from the inorganic through the organism back to the inorganic belongs every bit as much to the latter days of music as to the younger (and older) discipline of physiology. How could we possibly get on with fixed forms? Forms must change! That is a condition of life. The opposite is stagnation, pseudo-life, death. Form has no right to persist.[17]

The series of forms could, according to Marx, be infinite and their development was a product of reason. The system of forms, accordingly, appeared as applied logic in which fundamental forms functioned as the foundation and precondition for all secondary forms.[18] Marx based both his compositional pedagogy and his system of musical analysis on what he defined as the first two fundamental forms in music, the *Gang* (a continuation or transitional passage, approximately) and the *Satz* (a complete musical form, from phrase to entire movement). The Satz, containing the motifs of the musical work that was eventually developed, established the structural framework for the composition. The additions and expansions of harmonies and rhythms led to the construction of larger, more sophisticated forms up to sonata form, which for Marx, was the exemplar of musical organicism, and alone stood equal to the highest aesthetic demands of the age.[19] Notice that Marx's developmental system of forms did not reflect music history but rather a series of increasingly complex forms.

The agency of development lay in form itself. Form willed its expansion toward both greater complexity and compositional freedom. Musicologists have focused on the development itself and alternately interpreted Marx's understanding of the philosophy of musical form as either Hegelian or biological (Goethe's description of the development of plants). The musicologist Scott Burnham points out that neither of these developmental models successfully reflects music history or an individual's composition process and that they underemphasize the extent of Marx's pedagogical goals. Burnham instead argues that Marx believed it was the student's burgeoning artistic consciousness that grew and necessitated new musical forms.[20] Marx's arrangement of forms from simple to complex was both a pedagogical program—walking the student through the compositional process, building from simple forms to complex ones—and a means of freeing himself from analyzing musical forms in the traditional historical manner.[21] His teleology was based on formal coherence, with which to better analyze musical works and to articulate deeply idealist values.

Marx's formalist analytical system allowed him to validate and elevate the formal procedures and their use in the classical style, particularly in the music of Beethoven.[22] Certainly the bulk of his music-theoretical analyses employed Beethoven works as examples. Musicologist Sanna Pederson describes Marx's effort to promote the Beethoven canon by actively pushing for the performance of Beethoven works in their entirety. Indeed she credits Marx with changing performance practices to the presentation of multimovement works as wholes as well as the repeated performances of symphonies in a given performance season, especially Beethoven's *Pastoral* Symphony.[23]

This project was rooted in Marx's belief that Berlin audiences' appreciation of what would come to be known as the classical German repertoire was essential for understanding both their age and themselves.[24] Numerous scholars have noted the moral imperative that accompanied Marx's elevation of autonomous music to the idealist realm.[25] Whether it was a period of aesthetic elevation or decline, Marx explained that it was the people's "duty to devote to art the purest and noblest feelings, and to prepare ourselves for its service as diligently and carefully as possible."[26] Note the moral imperative.

Marx's system of musical form was idealist both philosophically and programmatically. Its pedagogical elements applied to the individual composition student and the German people. Form as developmental, bound to audiences, fulfills a classical, idealist social program. In his 1841 monograph, *Die alte Musiklehre im Streit mit unser Zeit* (The Old School of Music in Conflict with Our Times), Marx framed the role of autonomous music as potentially transcendent and capable of aiding humanity in overcoming a degenerate, material world.[27] For the process of composition was, according to Marx, none other than a "coming to consciousness of [the composer's] own inner being and artistic experiences."[28] The *spiritual* content of music, accessible through higher cultivation of those both creating and receiving music (so, the composers, performers, and listeners), elevated music to an art where it could benefit humankind.[29] The trick, then, was to cultivate an audience in which the purifying, spiritual side prevailed. This required listening in the proper way.

A formalist approach to analyzing music required a formalist approach to listening to music. The new performance practices promoted by Marx underscored a certain type of listening. His work consolidated a new attitude toward musical comprehension framed in terms of the formal elements. Such comprehension required the thoughtful and repeated listening to a piece in its entirety. Pederson argues that Marx's efforts, by incorporating the social responsibilities—indeed framing such responsibilities as a moral imperative—for promoting and preserving autonomous music, was instrumental in reifying its superiority among mid-nineteenth-century German audiences.[30] Eduard Hanslick, the leading music critic in Vienna, furthered this project.

A New Way of Listening

Eduard Hanslick was a Bohemia-born Viennese music critic, historian, and aesthetician. Hanslick dominated the Viennese music scene in the second half of the nineteenth century, reviewing works for the *Wiener Zeitung* and *Neue freie Presse*, adjudicating at competitions, and teaching as a professor of the history and aesthetics of music (one of the first such *ordinarius* professorships) at the University of Vienna. His work is generally considered to mark the beginning of a new aesthetic discourse in the nineteenth-century German-speaking world.

Hanslick was primarily a music critic but his 1854 treatise, *Vom Musikalisch-Schönen* (The Beautiful in Music) was a work of aesthetic theory, one of the very first to emphasize the role of attention or intellectual contemplation in the listener's experience of music. *Vom Musikalisch-Schönen* was intended as a refutation of the long-held belief that feelings were the substance of music. Rather than understanding music as the expression of the composer's emotions, or as a means by which to elicit emotions in the listener, or as an emotional exchange between the two individuals, Hanslick instead proposed that the nature of the beautiful in music is the specifically beautiful. The beautiful in music was form, both structure and musical idea (so, both structure and the content of the structure). Beauty in music had no object or aim beyond itself. Music had no goal of giving the listener pleasure nor was it in need of a subject introduced from without (for example, with the use of words).

Sonically moving forms (*tönend bewegte Formen*) were, according to Hanslick, the essence of music and functioned to express musical ideas.[31] These musical ideas were an object of intrinsic beauty independent of a complete and pure reproduction of them by a performer. It was the forms that the listener contemplated, delighting in the musical structures.[32] He did allow that music could sometimes possess the dynamic properties of emotional states (the increase and resolution of tension) but beyond *changes* in emotion, neither the expression nor representation of emotions could be considered the content of music or the subject of aesthetic analysis.[33] Musical aesthetics should, according to Hanslick, strive to be more exacting. To rival the imprecise aesthetics of emotion, he proposed an alternative of formalism in which the aesthetic value of music was sought in the formal musical structures.

Hanslick was inconsistent in his description of form. He gave the metaphorical example of an architectural arabesque, suggesting an expressionless structure, but then later described form as analogous to the musical idea itself.[34] It was this latter understanding that he employed more regularly. Form was the expression of the spirit (*Geist*) and it was the spirit itself. Form was the real substance, or subject, of music.[35] In this sense, Hanslick's understanding of form was similar to that of Marx: bound to its content.

The music critics and musicians of Hanslick's day heavily critiqued his efforts to replace the musical aesthetics of emotions or feelings with what they believed to be an analytical approach based on rigid and empty musical structures. Some went so far as to frame their critiques of Hanslick as noble defenses of music against a soulless positivism. It is revealing that the supporters of Hanslick were predominantly philosophers, writers, and natural scientists, including Robert Zimmerman, Hermann Lotze, Moritz Hauptmann, and Hermann Helmholtz.[36] I would venture that this was due to features common to both theirs and Hanslick's projects.

If form was the musical idea, then Hanslick's formalist aesthetics were a more precise and objective way of analyzing the musical idea. Further, thrusting toward a universalism that married the natural world to the aesthetic one, Hanslick understood the formal logic in music to be based on "certain elementary laws of nature which govern both the human organism and the phenomena of sound."[37] The "primordial law" of harmonic progression was the basis from which musical forms developed.[38]

Hanslick believed that it was the primordial or occult natural affinities between sounds that limited the music created by humans to rules of rhythm, melody, and harmony. These affinities, "though not demonstrable with scientific precision . . . [were] instinctively felt by every experienced ear, and the organic completeness and logic, or the absurdity and unnaturalness of a group of sounds, [were] intuitively known without the intervention of a definite conception as the standard of measure, the *tertium comparationis*."[39] Hanslick therefore connected humans to nature through form and logic. The forms, based on the logic of nature, were the building blocks of the musical composition.

To be sure, Hanslick did not believe that a scientific understanding of music, in the psychological or physiological sense, could contribute much to musical aesthetics, if anything at all.[40] In later editions of *Vom Musikalisch-Schönen* he did nod to Helmholtz's theory of consonance and dissonance and acknowledged the contributions made by advances in acoustics, anatomy, and physiology to the understanding of sound sensation. But these advances, Hanslick insisted, did nothing to illuminate the physiological process by which the sensation of sound became a feeling, an essential part of the musical aesthetic program.[41]

Hanslick acknowledged the growing effort in aesthetics to connect all art to natural processes and root it in the order of nature. And while he too sought points of intersection, Hanslick insisted that the natural world and the aesthetic one were separate. The musical system of humans did not exist in nature. Unlike the other arts, there was nothing beautiful in nature for music to model.[42] Of the fundamental features of music —rhythm, melody, and harmony—only rhythm existed in nature. He acknowledged that nature provided pitch that was used to construct tones and tuning systems, but that was nature's only contribution. The creation of music from sound required the "unconscious application of preexistent conceptions of quantity and proportion, through subtle processes of measuring and counting" by abstract humans; loosely, the mathematization of sound resulted in music.[43] The resulting formal structures of music were then quantifiable and therefore aesthetically analyzable.

For Hanslick, then, sound was a material, external phenomenon of nature. It was the human composer and performer who generated music.[44] The listener received it. This human product, music, did not exist independently in nature and was instead bound to a specific time and place, neither timeless nor universal. Music had a materiality, but both the phenomenon and its subject were the product of humanity. So, such stylistic structures as modulations or harmonic progressions, though beautiful for a time, become hackneyed and stale in as few as thirty years.[45] Beauty was historicist and it was mortal.[46] In the following chapter, we will see that Helmholtz struggled with the implications of such historicity of musical aesthetics.

The musicologist Carl Dahlhaus has argued that the seemingly paradoxical efforts by Hanslick to promote a universal definition of the beautiful in music that transcended history but that at the same time historicized his aesthetics were reconciled by his definition of form.[47] Because form for Hanslick was both musical structure and musical content or idea, it both transcended and was specific to history. This conception of form granted Hanslick great flexibility in his music criticism. And I would push this further by noting that it was this flexible historicity of form that allowed Hanslick to frame his aesthetic criticisms in developmental terms. Musical works (Western ones) he deemed poor were described as primitive, preverbal, and pathological. They lacked both universal, transcendental beauty (form) as well as a musical idea (also form), rendering them incongruent with the historical moment in which they were introduced.

For the most part *Vom Musikalisch-Schönen* focused on Hanslick's system of formalist musical aesthetics, but he also advanced a series of listening typologies. Just as there were primitive or pathological compositions and performances, Hanslick also critiqued pathological listening. Pathological listeners were in a state of passive receptivity, allowing the elemental features of the music to wash over and affect them in a vague, "supersensible" manner. Pathological listeners experienced music in a twilight state awash in sounding nullity.[48] There were also observant listeners, who, misguided by the musical aesthetics of emotion, fancied themselves to be musical because they could be emotional. These listeners devoted their attention to the superficial elements of a composition and sought only an abstract feeling or emotion. Hanslick derided these listeners as the lowest common denominators in the audience, in terms of ease of entertainment; they experienced music as if chloroformed.[49] He conceded that it was music's fleeting, ephemeral existence, as well as its distinctive beauty rooted in its form, that allowed it to be enjoyed with such a lack of reasoning.[50] Music was simultaneously imperative and indulgent.[51]

Hanslick distinguished these poor listening practices from the true method of listening, aesthetic listening. He defined aesthetic listening as the voluntary act of pure and reasoned contemplation complete with diligent attention and thoughtfulness. To listen with active but unflagging attention was "a truly aesthetic art in itself."[52] Attention itself became an aesthetic experience. This true aesthetic listening rewarded the listeners that followed and anticipated the composer's (formalist) intentions with intellectual satisfaction and delight. Significantly, this meant that it was the listeners' responsibility to locate beauty in music, not the composer's or performers' responsibility to present it to them.

This new importance if not primacy of the proper listener employing the proper listening advocated by both Marx and Hanslick marked the beginning of a trend in musical aesthetics and music criticism. Musicologists, music theorists, music aestheticians were redefining the act of listening. Musicologist Rose Subotnik holds Hanslick's formalism responsible for the priority and practice of "structural listening" in music criticism as well as the consequences of its prevalence.[53] Structural listening, Subotnik explains, is the sensory perceptual practice of following and comprehending a musical composition, with all its inner formal structures and relations, as it is realized in the listener's presence. She argues that this conception of structural listening assumed that the listener's mental framework was ideologically and culturally neutral, and required neither self-criticism nor awareness of individual or cultural biases. As such it was viewed as an objective equivalent to the methods of the pure sciences.[54] Subotnik sees the introduction of structural listening as central in the distancing of listeners—by condescendingly comparing other types of listening to drug intoxication, irrationality, and so on—from the sensuous experience of art music, an effect that is bound up with contemporary music today and contributing to its social irrelevance.[55]

Marx and Hanslick's sympathetic treatment of the structural listener indicates that the status of musical expertise in the act of listening was being renegotiated in the middle of the nineteenth century, part of a growing trend in music criticism and musical aesthetics. The musical experience of the individual who listened according to the proper rules of musical analysis was unquestioned. As long as the listening technique was correct (employed formalism), the sounds heard were legitimate.

Now recall that in addition to the redefinition of musical listening and musical analysis, tuning systems (equal temperament versus just intonation, for example) and even the Western tonal system itself had been destabilized by changing technology and Europe's exposure to non-Western music. It was a sonic world in which almost anything could be heard. It was to this world of sonic anarchy that the early musicologist Hugo Riemann first proposed his dualist system of music theory based on the existence of (nearly) inaudible undertones.

A Newer Way of Listening

The brief period of engagement by natural scientists with Riemann's experimental work connects this chapter to the larger themes of this book. This intellectual exchange was possible for two intertwined reasons. First, the individual, subjective experience was becoming an increasingly legitimate object of interest in both discussions of music reception and psychophysical experimentation. Riemann's detractors could not contest that he might very well have heard something. They could only critique the universality of his claimed experience. Second, there was an increasing awareness that the musical sounds individuals were exposed to were changing as well. New tuning systems and new non-Western music had destabilized the established structures of musical aesthetics. This tonal destabilization, combined with the growing acceptance of individual, subjective experience of sound, created an opportunity for Riemann to try to support his musical aesthetics through a little psychophysical experimentation.

Riemann, leader of the generation that followed Marx and Hanslick and often described as the "father of musicology," desperately sought to make order of this sonically unstable world. His painstakingly researched writings covered every branch of musicology and, in a departure from the music-theoretical works of Marx and Hanslick, often drew on the natural sciences to buttress his claims. Riemann reinforced the validity of his theoretical system of harmony through his historical analysis of classical masterpieces and then further disseminated his theories through pedagogical texts. His theoretical works engaged academics and provided guidance for contemporary musicians and composers, including Max Reger and later Arnold Schoenberg.

In 1871, following his military obligation during the Franco-Prussian campaign, Riemann pursued music at Leipzig Conservatory. Riemann's dissertation, *Über das Musikalische Hören,* in which he first presented his ideas on harmonic dualism, was initially rejected at Leipzig. Later, after he secured the support of Hermann Lotze, it was accepted at Göttingen. From 1873 to 1878 Riemann was a conductor and teacher in Bielefeld. Then for the next twenty-five years he taught music theory and piano performance in Leipzig, in Bromberg, and at the Hamburg Conservatory, then at the Wiesbaden Conservatory. Riemann also wrote prolifically on musical aesthetics.[56] He saw the project of musical aesthetics as a joint one among artists, natural scientists, philosophers, historians, and philologists alike, and read widely, including the writings of Fechner and Helmholtz.[57] He highlighted Fechner and Helmholtz's exacting efforts to prove the mechanical processes associated with tone sensation, and he noted that Fechner's psychophysical correlation between tone frequency and the aesthetics of musical harmony and melody was only one hypothesis of many.[58]

From his very earliest writings, Riemann understood the ear to be an organ of reception. Indeed, the title of the work in its 1873 dissertation form was *Über das musikalische Hören* (On Musical Listening).[59] The text began with a discussion of how the ear analyzed sound, then moved through the sensory perception of consonance and dissonance to the tonality of major and minor keys to harmonious phrasing to conclude with a discussion of Rameau's theory of grounded bass.[60] The work was structured philogenetically, as both the experience of the individual receiving sound and eventually perceiving music *and* as a history of the development of the Western music system generally. Focusing on the psychological and physiological (dare we say psychophysiological?) basis of music allowed Riemann to toggle between discussions of the individual experience of music and musical aesthetics generally.

Riemann's interest in the listener and the process of listening existed from his earliest works. Philosopher Trevor Pearce has shown that in both *Musikalische Logik* and *Musikalische Syntaxis,* Riemann presented an understanding of musical hearing as an activity of the mind and suggested that this musical logic was the basis for musical harmony.[61] In the preface to his 1888 collection of three lectures, *Wie hören wir Musik?* (How Do We Listen to Music?), Riemann identified the clash of musical aesthetic positions between Hanslick's formalism and the belief that music was capable of expressing poetic ideas musically.[62] Riemann also identified a third approach that combines and perfects the first two: *his* analysis through a distinction between the elementary, the formal, and the characteristic elements of music. These elements, in turn, informed an understanding that the listener perceived music in two rather different ways, depending on the type of music. In the first, more often in absolute music, in which the formal dominates the elementary, music was ultimately felt as the manifestation of one's own will, subjectivated. In the second form, which occurred more with program music, music was objectivated by the imagination. Riemann's ultimate goal was to illuminate the process by which music was produced in the mind of the composer as well as how the effect of the music on the listener necessarily followed.[63]

Though he devoted significant ink to trying to wedge Western music theory as well as his beloved symmetry into this theory of a subjectivation-objectivation process, *Wie hören wir Musik?* was ultimately about listening and the listener. Indeed Riemann opened the section on harmony and rhythm with a quote from Fechner's *Vorschule der Aesthetik*, paraphrased toward his own ends that the impulse to play was rooted in a desire to give purely subjective pleasure to the subject.[64] The goal of a performance was to engage the listener.

In turn, listening was an active process for Riemann. His understanding of listening elaborated on both Marx and Hanslick's models. Like Marx, he believed a musical work to be a complex whole consisting of simple forms. Aesthetic analysis required grasping these passing structures as the music unfolded. The musical whole would not be apparent to the listener until the piece was completed, thus requiring careful and sophisticated listening.

Further, in the higher, art forms of music there was no possibility of comprehension through the passive surrender to impressions only. Instead, just as Hanslick would recommend, active participatory listening was required.[65] This did not necessarily require special technical knowledge, Riemann continued, though musical practice and education were indispensable for a full listening experience.[66] Pearce argues that Riemann's conception of active, participatory listening drew on the Kant-inspired views of his educators and contemporaries.[67] Pearce explains that Riemann understood musical hearing to be a form of active logic and therefore an act in which the hearing and analysis of music were unified. To hear music was to also analyze it. I think we can understand Riemann's conception of musical hearing to be psychophysical.

To analyze music, Riemann promoted a system of harmonic dualism, a system of symmetry in which major and minor chords (specifically, the triad) were understood as inversions of each other. Riemann's reading of the physicist Arthur von Oettingen's 1866 treatise, *Harmoniesystem in dualer Entwickelung: Studien zur Theorie der Musik*, was perhaps the most significant source of inspiration for his own work on harmonic dualism.[68] Oettingen was a professor of physics, the president of a musical association, and the director of the amateur orchestra in Dorpat.[69] Apparently, the great strength of Dorpat physicists during the mid-nineteenth century was meteorology. Oettingen was no different. He gained international attention through his development of the anemograph, a means of determining wind speed and direction.[70]

Oettingen's *Harmoniesystem in dualer Entwickelung* was an application of his expertise with pressure waves to acoustics, framed as a response to Helmholtz's *Tonempfindungen*, published three years earlier. Oettingen took issue with Helmholtz's definition of consonance and dissonance. For Helmholtz the criterion for a tone or a combination of tones to be consonant or dissonant was the sounding of beats. Beats indicated dissonance. The presence of harmonic overtones indicated consonance. Overtones were consonant because no beats were heard in the overtone series. Oettingen pointed out that even in a single tone the higher overtones could create beats with each other, and that all tones contained features of both consonance and dissonance as defined by Helmholtz. Thus, according to Oettingen, overtones could not, as Helmholtz had asserted, be the basis for the theory of music.[71]

Oettingen instead proposed a dualistic harmonic system in which the major and minor triads were mirror inversions of each other.[72] The major triad—for example, C-E-G—was mirrored by the minor triad C-A^{b}-F downward. This could be extended further to the major and minor key systems in which, for example, the tonic major scale of C (C-D-E-F-G-A-B) was mirrored by the phonic scale read downward (E-D-C-B-A-G-F-E, the A-minor scale descending from its dominant).

Riemann aligned himself with Oettingen against Helmholtz.[73] He similarly believed the one-sided definition of consonance to be a significant weak point of Helmholtz's physiological basis for music theory. Riemann further believed that Helmholtz, in supporting such a system while advocating just temperament, had been "caught in his own snare." A minor chord sounded in just intonation with perfectly pure intervals would, by Helmholtz's criteria, appear to be so definitively dissonant that musicians would protest.[74]

This did not, however, stop Riemann from borrowing elements of Helmholtz's overtone theory of consonance as needed for his own theory of harmonic dualism.[75] In the 1874 publication of his dissertation (*Musikalische Logik*) Riemann presented the foundations of his own theory of harmonic dualism, much like Oettingen's, in which the mirroring of the major and minor key systems was rooted in a mirroring of harmonic overtones and harmonic undertones (see figure 2.1).[76] Major consonance consisted of the overtone series, the harmonics ascending from the principal tone—Riemann was in agreement with Helmholtz on this point.[77] Minor consonance, according to Riemann, was the undertone series, the harmonics descending from the principal tone.[78] Every tone emitted a series of both undertones and overtones. In an individual's perception of a single tone, one series would dominate and the other was erased.

Riemann's "Grid of Harmonic Relations"—a means of navigating the twelve keys of the Western tonal system and another appropriation from Oettingen—complemented his theory of harmonic dualism. One can see in figure 2.2 that the components, centered around C, are related horizontally by fifths, thus delimiting the circle of fifths of the twelve keys. One could, for example, construct a C-major chord by selecting the box above (E, the third) and the box to the right (G, the major fifth). The diagonal relation between the E and the G box was a major third. Or, alternatively, one could construct a C-minor chord by selecting the box below (A^{b}, the

Figure 2.1
The overtones and undertones for middle C. Hugo Riemann, *Musikalische Logik*, 2.

5	4	3	2	1	0	1	2	3	4	
Cis	Gis	dis	ais	eis'	his'	fisis"	cisis"'	gisis"'	disis""	3
,A	E	H	fis	cis'	gis'	dis"	ais"	eis"'	his"'	2
,F	C	G	d	a	e'	h'	fis"	cis"'	gis"'	1
,Des	,As	Es	B	f	c'	g'	d"	a"	e"'	0
,,Bb	,Fes	Ces	Ges	des	as	es'	b'	f'	c"'	1
,,G*bb*	,D*bb*	,A*bb*	E*bb*	Bb	fes	ces'	ges'	des"	as"	2
,,E*bbb*	,,B*bb*	,Feses	Ceses	Geses	deses	ases	eses'	heses'	fes"	3
5	4	3	2	1	0	1	2	3	4	

Figure 2.2
The Grid of Harmonic Relations. Hugo Riemann, *Musikalische Logik*, 29.

third) and the box to the left (F, the minor fifth). The interval between the A^{b} box and the F box was a minor third. Based on the assumption that all chord progressions and key modulations could be reduced to fifth and third relations, harmonic maneuvers were mapped out spatially and were nearly infinite in possibility. The grid was a demonstration of the full harmonic potential of Western music, what was compositionally possible.[79]

Riemann's Grid of Harmonic Relations was complemented by his taxonomy of harmonic function based on the three pillars of harmony—tonic, dominant, subdominant chords—derived from Rameau. This taxonomy of harmonic function related chords to their tonic in an effort to delimit the extent of harmonic successions possible within a given key, what was compositionally permissible. This two-part theory of harmony, involving dualism and function, allowed Riemann to release music theory from its dependency on the fifth and fully realize the potential provided by equal temperament. It also indicated the tensions that arose when attempting to reconcile a universalistic conception of harmony (the dualistic Grid of Harmonic Relations) with Western harmonic functions, and the implication of these tensions for music composition.

A New Sound?

So, those harmonic undertones. Present-day acoustics tell us that undertones as Riemann conceived of them—a descending series of partial tones mirroring the harmonic overtone series—do not exist. Even in Riemann's time there was little reason to believe that undertones existed.[80] And yet, in part because the legitimacy of his theory of harmonic dualism depended on their existence, Riemann went to great lengths to claim that they did.

The harmonic undertone was central to Riemann's theory of harmonic dualism. His explanation for what exactly the undertone was changed, evolving loosely, from a psychophysiological to a physical phenomenon. Riemann buttressed each of his explanatory models with what he believed to be empirical support. The music community for the most part enthusiastically accepted Riemann's work. Those from the natural sciences were, however, more suspicious. Though willing to acknowledge that Riemann may very well have heard *something*—again, due to the instability of tuning both within single instruments and between many, any individual's auditory experience was valid—they could not allow Riemann to ignore the laws of physics. Riemann's eventual transition away from natural scientific explanations for his undertones to more historical ones, and even his reformulation of undertones into the framework of Carl Stumpf's principle of tone fusion, can be understood as a yielding to the natural sciences and an abandonment of his efforts to take on the laws of nature.

Riemann likely originally came to his belief in the undertones from a music theory perspective, long before he saw the need for physical or psychological defenses of their existence. Indeed, in his very first publication, *Musikalische Logik: Hauptzüge der physiologischen und psychologischen Begründung unseres Musiksystems* (Musical Logic: Main Features of the Physiological and Psychological Justifications for Our Music Systems), the 1874 publication of his dissertation, he demonstrated the existence of undertones through an analysis of musical works. (Note the circularity of using music to demonstrate the existence of physical phenomena that are the basis of Western musical aesthetics.) In figure 2.3, see Riemann's diagram of the overtones and undertones generated (*Obertongeschlecht* and *Untertongeschlecht*) by the then well-known melody, "Ach, wenn du wärst mein eigen."[81]

Riemann began *Musikalische Logik* with the claim that the work was an attempt to combine the work of Jean-Phillipe Rameau, Moritz Hauptmann,

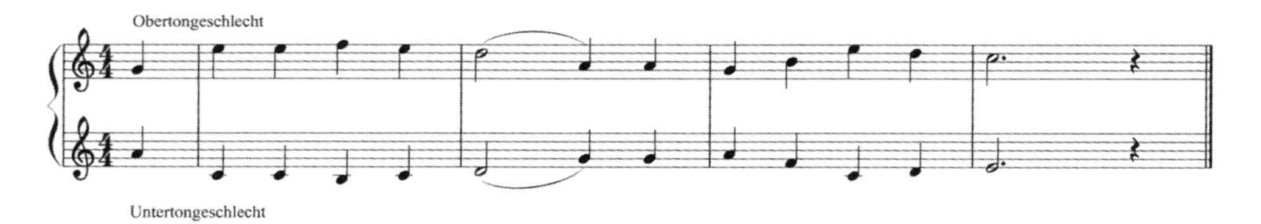

Figure 2.3
Overtones and undertones of "Ach, wenn du wärst mein eigen." Hugo Riemann, *Musikalische Logik*, 46.

Arthur von Oettingen, and Hermann Helmholtz. He explained that the hypothesis he advanced was nothing more than an articulation of the consequences of Helmholtz's evidence concerning the internal construction of the ear and the analysis of sound.[82] It is noteworthy that Riemann considered himself up to the task of reconciling the acoustical research of these celebrated experimentalists.

Borrowing from a model employed by both Helmholtz and Ernst Mach, Riemann described the physiology of the ear as analogous to a piano. On the piano, the struck string and those corresponding to the harmonics of the struck string vibrated (the latter sympathetically). Similarly, the cortical fibers of the inner ear that correspond to both the sounded tone and those of the sounded tone's harmonics—both overtones and undertones—and those corresponding to the harmonic's harmonics all vibrated.[83] With the significant exception of additional fibers sympathetically vibrating in response to the undertones, this was Helmholtz's physiological model of the ear. Riemann grafted his theory of undertones onto it.

According to the piano model, the piano strings correlating with overtones and undertones that vibrated in response to the struck tone only did so once the string of the struck tone vibrated. This suggests Riemann believed that only the sounded tone would enter the ear and that the harmonic overtones and undertones would only occur and subsequently sound their corresponding cortical fibers within the ear. The implication was that harmonic overtones and undertones were not necessarily a physical phenomenon but a *physiological* one.

Riemann insisted that the undertone did sensorially exist, if only in the cortical fibers of the ear and even if the individual was not conscious of it.[84] In the appendix to his translation of Helmholtz's *Tonempfindungen*

Alexander Ellis acknowledged Riemann's physiological theory, although he was quick to point out that it was not observable in the ear. But Ellis continued further to state that no objection could be made to Riemann's work because the appendages of the basilar membrane of the ear could very well impede the formation of tones with nodes.[85]

In his 1875 pamphlet, *Die objective Existenz der Untertöne in der Schallwelle* (The Objective Existence of Undertones in Sound Waves), Riemann advanced a physical explanation for the undertone. His explanation was an expansion on known acoustics. Just as the overtones of a fundamental tone sounding at frequency *x* will sympathetically vibrate at 1/2 *x* and 1/3 *x* and 1/4 *x*, Riemann argued that so also must "the entire undertone row result . . . in evenly diminishing strength toward the depths."[86] So, for the fundamental C, where the overtone series ascended through the pitches c, g, c′, e′, g′, b^{b}′, c″, e″, etc., Riemann believed there was also a descending series of A^{b}, B^{b}, c, d, f, a^{b}′, c′, f′, c″, etc. He believed that this could be proven and then proceeded to try.[87]

Though Riemann's work on harmonic dualism was initially well received by most of the music world, there were critics who questioned the existence of the undertones he claimed to hear that were so critical to his theory. Riemann replied to them in 1877: "However this may be, and if all the authorities in the world appear and say: 'We hear nothing', I must still say to you: 'I hear something, and indeed something very significant.'"[88] And yet the replication of Riemann's work appeared to be limited to his own personal piano (an Ernst Irmler).[89] Nobody else could hear them. He acknowledged that even Oettingen could not hear undertones on his piano (a Breitkopf und Härtel).[90] And yet Riemann considered his individual, subjective experience of sound valid and just as legitimate as the silence heard by "all the authorities in the world," even Oettingen, whom he had followed so closely. Riemann's confidence was a testament to the freedom permitted by the instability of the tonal world as well as Marx and Hanslick's efforts to redefine the act of listening.

Recall Karl von Schafhäutl, professor of geology, mining, and metallurgy at the University of Munich, who had also worked on acoustics and musical instrument design, pipe organs in particular. In an 1878 contribution to *Allgemeine Musikalische Zeitung*, Schafhäutl pointed out that on the border of "musical audibility" (*musikalischen Hörbarkeit*) it was quite possible to hear whatever one wants to hear.[91] He recounted an anecdote about a conductor asking the horns to play still more *pianissimo* when they had in fact already completely stopped playing the passage out of frustration. Or take Beethoven, Schafhäutl continued, who intended his audiences to experience particular tones and effects though he was himself completely deaf.[92]

Although Riemann may have heard his undertones by the same processes that allowed Beethoven to compose symphonies in spite of deafness, Schafhäutl insisted that the existence of undertones conflicted with physical law. He attacked: "One is trying to force Nature to sound undertones, which cannot exist according to the laws of mechanics, and the most recent dualist Dr. Hugo Riemann tells us that he and nobody else, save perhaps for Aristotle two thousand years ago, has heard these undertones, which alone are supposed to explain the consonance of the minor triad."[93]

Riemann's claims of empirically demonstrating the existence of undertones, though acknowledged, were dismissed by natural science as physically impossible. Schafhäutl maintained that Riemann's undertones were nothing more than wishful hearing, an attempt to build a theory on the apparition of tones (*Erscheinung von Tönen*) that nobody could hear.[94]

Alexander Ellis was more conciliatory. In the commentary of his 1885 translation of Helmholtz's *Tonempfindungen*, Ellis acknowledged the indispensable contribution of Riemann's harmonic dualism to composition theory. And yet Ellis maintained that he was unable to convince himself that undertones sounded when a tone was struck on a piano with its dampers lifted. He suggested that perhaps a particularly violent striking of tone on the piano shook the entire instrument, causing the lower strings to vibrate. Or that Riemann was deceived by the sympathetic vibration of upper partials (overtones) reinforcing the struck fundamental tone: Riemann heard wrong.[95]

Efforts by acoustical instrument makers at the end of the nineteenth century to develop ultrasonic tuning forks may have given natural scientists responding to Riemann some reason to pause. Historian David Pantalony documents the small scandal that erupted around the integrity of the Appun workshop tuning forks and Rudolph Koenig's resulting comprehensive study of inaudible tones (up to 90,000 cycles per second).[96] Still, those forks generated sound waves whose existences were confirmed by tests with vibrating plates, steel bars, and whistles; they were simply outside the range of human hearing, ultrasonic but real.

In the end, the response from the natural sciences to Riemann's work was that whatever Riemann believed he had experimentally observed were not undertones because undertones were not physically possible. It was acknowledged that psychological processes might have allowed Riemann to perceive certain sounds and that his physiological explanation might have had some validity, but both of these possibilities had to defer to the laws of mechanics. Undertones did not exist in the physical world. Even Helmholtz, in a later edition of *Tonempfindungen*, suggested that Riemann had been deceived (*getäuscht*).[97]

Riemann for the most part deflected the attacks of music theorists by simply ignoring them.[98] But he continued to adjust his empirical defense of undertones and attempted to shift the burden of proof to his critics in the natural sciences. In 1882 he insisted that a vibrating tuning fork placed on a sounding board sounded "*not the fundamental tone* of the plate or fork, but its under octave, under twelfth, even the under double octave, under seventeenth, or other low undertones."[99] Riemann reiterated that every sounding tone generated both an overtone series and an undertone series and that his diligent collecting of information revealed nothing inconsistent with this hypothesis.[100]

Besides, the system of harmonic dualism, dependent on the existence of harmonic undertones, was an effective way to analyze music. In his discussion of Brahms's *Doppelkonzert*, the 1889 article "Einige seltsame Noten bei Brahms und anderen," Riemann showed that his system of harmonic dualism could aid in the analysis of complicated passages—in particular, better illuminating key changes. Correspondingly, the harmonic undertones remained apparent as well and Riemann included a demonstration of *Doppelkonzert*'s dualistic qualities.[101] Riemann explained that in the finale of the piece (see figure 2.4) Brahms introduced the minor-subdominant (*Moll-Unterdominante*) of the major key (*Durtonart*) such that a minor-Major passage developed. Riemann diagrammed this transition from subdominant to tonic to dominant keys according to Hauptmann's "*Molldurtonart*" chord system (see figure 2.5).

Riemann then explained that by inverting (*übersetzen*) Brahms's passage from minor-major to major-minor (see figure 2.6), one could confirm that an analogous *Durmolltonart* chord system occurred (see figure 2.7). A coherent analysis of the transition from minor to major in Brahms's original passage and from major to minor in its inverted form was successfully

Figure 2.4
Brahms, *Doppelkonzert* passage, op. 102, Finale. Hugo Riemann, "Einige seltsame Noten bei Brahms und anderen," 109.

Unterdominante Oberdominante

f as c e g h d

Tonika

Figure 2.5
Hauptmann's "*Molldurtonart*" chord system (transposed to C major) for the Brahms *Doppelkonzert* passage. Hugo Riemann, "Einige seltsame Noten bei Brahms und anderen," 110.

Figure 2.6
The inverted Brahms *Doppelkonzert* passage. Hugo Riemann, "Einige seltsame Noten bei Brahms und anderen," 112.

achieved through the application of the dualist harmonic system. It corroborated the dualist prediction that fifths were likewise the under-fifths of the fundamental tone of the minor key.[102] In this way, Riemann's analysis demonstrated that a very tangible unity of understanding was possible through his two-part (dualism and function) theory of harmony.

By the 1890s Riemann had retreated from his claim of hearing undertones. He maintained, however, that this did not mean undertones did not exist. Backing away from earlier empirical defenses of the existence of undertones—"I can hear something, something very distinct"—Riemann instead turned to wave dynamic theory to maintain the undertones' objec-

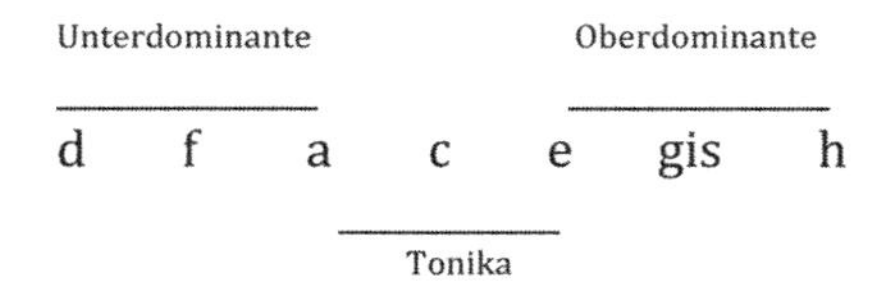

Figure 2.7
The *Durmolltonart* chord system for the inverted Brahms *Doppelkonzert* passage. Hugo Riemann, "Einige seltsame Noten bei Brahms und anderen," 112.

tive reality. In his encyclopedic *Musiklexicon* he claimed by proof "of a scientific character" (*wissenschaftlichen Beweis*) to have shown that "each tone, out of necessity, produces the entire series of undertones; but each, according to its ordinal number, so many times; the second, twice, the third, three times, etc., but proceeding in such a way that by interference, they must neutralize one another."[103] In his 1891 publication, *Katechismus der Akustik: Musikwissenschaft* (Catechism of Acoustics: Musicology), Riemann explained that though the undertones physically existed, the individual's experience in which he or she "nevertheless does not hear the tone *c* if *g* rings out, is explained by the law of interference."[104] Wave interference prevented the ear from hearing the undertones because they were canceled out before they ever reached the ear.

If the undertones did reach the ear, Riemann continued, the perceptual process of "tone fusion" (*Klangverschmelzung*) would obscure the experience of overtones and undertones.[105] He borrowed this concept of tone fusion from Carl Stumpf, who had first presented it in his *Tonpsychologie*, a two-volume work published between 1883 and 1890. Tone fusion was the perception of two tones fused as one entity and was conditioned by "the natures of tones or the brain processes on which they are based."[106] For Stumpf the likelihood of two tones perceived as fused was the basis for harmonic consonance in music. This was an explicit refutation of Helmholtz's overtone-based theory of consonance by Stumpf. It is therefore interesting that Riemann employed the work of both Helmholtz and Stumpf to support his claims that the undertones existed.[107] Features of the physical sound object as conceived by Helmholtz (that wave interference canceled out the undertones) and the psychophysiological process of tone fusion by Stumpf (that the perception of consonance as fused tones would obscure any perception of undertones) explained why Riemann's undertones went unheard.

Then, Riemann stopped discussing undertones entirely. His interests turned elsewhere. His decades-long effort to establish the existence of undertones, despite a complete lack of training in the natural sciences, by advancing physiological, physical, and psychological explanations for the sensation and perception of undertones is at first baffling. But it is ultimately indicative of the extensive room in which psychophysical studies of sound sensation could move.[108] Ultimately, Riemann's use of natural science to demonstrate harmonic overtones and buttress his theory of harmonic dualism was possible because of several factors: first, the upheaval of the tonal landscape caused by shifts in tuning practices and the popular introduction of non-Western music; second, Marx and Hanslick's renegotiation of the very definition of listening; and third, Riemann's own understanding of musical listening as an active, analytical process.

A New Discipline

Although Riemann eventually abandoned his undertones—or at least stopped defending their existence—he was successful in his larger project of unifying the new discipline of musicology around a theoretical base.[109] In 1895 he returned to Leipzig and in 1901 was granted an *extraordinarius* professorship in music. He was made director of the Collegium Musicum in 1908 and then of the newly established Forschungsinstitut für Musikwissenschaft in 1914, confirming his critical importance in the new discipline of musicology.

The usual indicators of a disciplinary status cannot be seen until the end of the nineteenth century for the scholarly study of musical aesthetics.[110] Although Eduard Hanslick was granted a chair in music history and aesthetics at the University of Vienna in 1861, the first ordinarius chair in Musikwissenschaft, also at the University of Vienna, was not established until 1870. The second chair was established in Strasbourg in 1897, then in Berlin in 1904 and Munich in 1909.

In terms of academic journals, one could look to Robert Schumann's *Neue Zeitschrift für Musik* in Leipzig, established in 1834, as part of an effort to combat what he perceived as the decline of German music in the vacuum following Beethoven's death. The *Allgemeine Musikalische Zeitung*, a highly esteemed weekly music periodical that ran from 1798 to 1848, was revived in 1863 by the Leipzig publisher Breitkopf und Härtel and ran as a monthly journal until 1882, although it was never commercially successful. Marx's *Berliner allgemeine Musikalische Zeitung* only lasted seven years, from 1824 to 1831. These two journals were devoted mainly to music criticism, concert reviews, book reviews, and discussions of music theory. The first journal explicitly dedicated to the scholarly study of music was the *Vierteljahrsschrift für Musik*, established in 1885 and published for nine years. In 1898 Carl Stumpf established the *Beiträge zur Akustik und Musikwissenschaft*, which lasted until 1924. Both the *Vierteljahrsschrift für Musikwissenschaft* and the *Beiträge zur Akustik und Musikwissenschaft* published articles by musicologists as well as natural scientists.[111]

In addition to the growing institutional structure of academic musicology as well as the analytical programs presented by Marx, Hanslick, Riemann, and others, a vigorous discourse about the origins of music also developed at the end of the nineteenth century. Not too surprisingly, the young discipline became preoccupied with addressing the very issue of how music evolved the way it did, hitherto ignored by such evolutionary theorists as Darwin, Spencer, or Bücher. The work generated by this preoccupation, however, was not evolutionary in the vein of Darwin, Spencer, or Bücher (the best example of a superficial application of Darwinian evolution to musical aesthetics can be found in Eduard Kulke's work, discussed in chapter 4). It was historical.

Rehding argues that the musicologists' efforts to establish a developmental history for music maintained a philosophy of origins in which longevity legitimized the origin. This was not evolutionary theory. The origin shed its pastness; it was the true origin because it was still present. This was reinforced by a vague and conflated definition of music that equated musical parameters with Western tonal music. Thus, musicologists' searches for origins were not so much concerned with points of departure as with determining and then maintaining original practices.[112] The origin was identified through the tradition being legitimized.

Numerous musicologists therefore struggled to explain the frequent lack of Western tonality in other times and places. Some argued that the ability to comprehend harmony was innate and simply did not exist in some racial types. Or that there existed a greater universal musical logic of which Western tonality was only one possible manifestation. Others viewed Western tonality as product of the historical and cultural variability of hearing. All of these perspectives acknowledged a physical and physiological basis for tonality.[113]

Riemann's music history did not conform to the basic tenets of the German musical historicism that tended to place all things German at the end of a single progressive development of civilization and culture. Riemann instead sought to "make recognizable that which is a primordial law in all ages."[114] Both historical and national differences in style were merely superficial manifestations of a greater universal musical logic. He rejected the suggestions of comparative musicologists that non-Western music did not comply with his notions of harmonic function. Riemann countered these criticisms by demonstrating the universal validity of his theories with such efforts as a collection of original Chinese and Japanese melodies arranged for Western instruments like the violin and piano.[115] This universal musical logic did not, however, preclude development. Riemann in fact believed that no other art could demonstrate "a continually progressive development like that found in music."[116] He claimed that cultures gradually gained awareness of the immutable laws of musical logic. Not too surprisingly, Riemann's genealogy of music theory realized his own harmonic dualism.[117]

Still, this genealogy of music theory was a historical framework, organized by universal musical logic, not laws of acoustics or evolutionary theory. Despite ever-shifting supportive arguments, Riemann had, from the very beginning of his career, consistently maintained a firm stance on the existence of harmonic undertones and an even stricter position on the universal truth of harmonic dualism. His Grid of Harmonic Relations, taxonomy of harmonic functions, and search for immutable laws of universal musical logic all reflect his preoccupation with order. This preoccupation extended beyond music theory to politics. Riemann had been known to launch the barb of "Social Democrat" at the theorists of the chromatic movement (Melchior Sachs, Albert Hahn, and Heinrich Vincent) for their lack of music-theoretical responsibility.[118]

Riemann was seeking order in an unstable musical world. He believed in a universal musical logic that was gradually realized by all cultures over time. This logic could be represented in an abstract form. Riemann described the task of art theory as the investigation of "the natural laws, which consciously or unconsciously rule the creation of the art and present them in a system of logically coherent rules."[119] Elaborating on this point in 1904, he stated that the task of Musikwissenschaft was to explain "the spiritual and expressive nature of the primitive elements of all musical experience . . . to ascertain the physical properties of tones and mechanical conditions governing their creation."[120] The discipline of musicology at the end of the nineteenth century sought to build large frameworks for understanding music—frameworks based on laws, hopefully some of which were universal physical laws, but only if they didn't get in the way.

Conclusion

For a brief moment in the mid-nineteenth century, new formalist theories exemplified in the work of Marx and Hanslick that reframed the listeners' experience of music as fraught with meaning and value, combined with the instability of the music world—of the *sound* world—to allow for undertones to be heard. So, fleetingly, Riemann could convince others through various proofs "of a scientific nature" that the undertones he heard did in fact exist. This brief intersection of psychophysics and musicology is an illustration of the two not-so-separate worlds, plagued by common issues presented by destabilized tuning systems and the introduction of altogether new non-Western sounds, unified in common inquiry about the nature of listening and engaged with each other's work. The tonal destabilization and growing acceptance of individual, subjective experience of sound created the opportunity—the need, even—for new frameworks of understanding. Riemann's abandonment of natural scientific explanations for his musical aesthetics and subsequent pursuit of music-theoretical and historical methods of studying music exemplify the development of musicology. Accordingly, these conditions that made Riemann's undertones audible also made modern musicology possible.

The unstable tonal world created by the changing music criticism and tuning systems, and the introduction of non-Western music, allowed Riemann to venture into the natural scientific world for explanations supporting his theory of harmonic dualism. But these universal physiological, psychophysical laws ultimately failed to explain his individual, subjective experience of harmonic undertones. Riemann devoted the latter part of his career to historical studies of music and theories of musical development. He used his examination of historical works in combination with his aesthetic program to determine the original, primitive elements of music, hoping to unify acoustics, aesthetics, and history. Riemann's transition to a search for the origins of music can be seen as emblematic of the maturation of the new discipline of musicology (Musikwissenschaft), with which the maturation of his career coincided.[121] The sounds of the nineteenth century could not be stabilized by universal scientific laws alone, but musicology, through music-theoretical analyses and historical studies of music, could provide a new framework with which to understand the changing world of music.

Riemann's intellectual arc, from a psychophysical understanding of listening and a struggle to reconcile universal systems or laws with an individual experience of sound to an eventual turn toward history, bears similarities to several of the psychophysicists examined in this book. His career is also emblematic of the discipline of musicology. He began with a concept of the listening process that was sophisticated (arguably rooted in Kantian logical theory and Stumpf's psychical functions) but, ultimately, eschewed natural scientific support and refocused on how individuals listened to and analyzed music as well as placing musical systems within a developmental framework. Riemann's presence is felt in modern musicology, which remains structured around formal analysis, reception theory, and historical approaches to musical aesthetics.

Though musicologists and psychophysicists struggled with the same difficulties of reconciling individual experience with their respective searches for universal systems, in the end, musicology firmly prioritized music-theoretical systems and a historical approach over the psychophysical. Riemann tested the waters of science and decided they were too cold. The larger implication for this work is that the cultural and intellectual exchange was more important for the psychophysical studies of sound sensation than for musical aesthetics. Musicologists had less to gain from the world of science, at least when it undermined the symmetry of their harmonic systems. It is perhaps the rare example of science valuing culture more than culture valued science.

At the same time, Riemann made a career out of systematizing musical thought, defining the discipline of musicology broadly to include the scholarly study of aesthetics, music history, acoustics, and the listening experience and locating points of intersection between these disparate subfields. It was in this way, Rehding argues, that Riemann was able to define music as a stable, knowable object *worthy* of scientific scrutiny.[122] For the purposes of this book, I would highlight that this was how Riemann was able to make the experience of listening—sound sensation—an object of investigative interest to those interested in psychophysical questions, to those interested in musical aesthetic questions, and to those interested in both.

3 Sound Materialized and Music Reconciled: Hermann Helmholtz

I always care much more for music when I am playing it myself.
—Hermann Helmholtz, letter to his parents[1]

I may as well make known, as a remarkable discovery in psychical physics, that the modes in which we express ourselves musically, that is, the major and minor scales, though in theory series of sounds bearing a fixed pitch in relation to one another, are in practice tempered by every musician just as the proportion of the human figure are tempered by a sculptor.
—George Bernard Shaw, *Music in London, 1890–1894*[2]

Hermann Helmholtz was a master of physiology, physics, and mathematics. His dominance in all three disciplines as well as his more thoughtful, philosophical writings were cherished and promoted as emblematic of the successes of the maturing German university system. As a young scientist in the 1840s, Helmholtz was one of the founding members of the Berlin Physical Society. The reformist goals of this group—an emphasis on an empirical approach to knowledge in German academia and a liberalized and modernized Prussian state—were incorporated into much of Helmholtz's science.[3] In his work on sound sensation in particular, much of it compiled in his 1863 opus, *Die Lehre von den Tonempfindungen als physiologische Grundlage für die Theorie der Musik* (On the Sensations of Tone as a Physiological Basis for the Theory of Music), the strength of his own liberal, upper-middle-class tastes and sensibilities was manifest.

Tonempfindungen was enormously popular, respected and studied by physiologists and physicists as well as music theorists and composers. Offering objective, experimental support, Helmholtz presented a physical definition of musical tone and a physiological explanation for the sensation of it. He sought natural laws of sound sensation that explained current musical systems.

And yet, by this time in the middle of the nineteenth century, there was a growing awareness that these musical systems—musical aesthetics—varied between cultures and over time. Although Helmholtz was careful not to tread into the realm of perception theory, instead restricting his claims to sound sensation, he certainly acknowledged such cultural and historical variation. So how was he able to reconcile his universal, lawlike understanding of sound sensation with a belief that musical aesthetics were culturally and historically bound?

I argue that there were two main avenues by which Helmholtz was able to achieve this reconciliation. First, it was accomplished through the status Helmholtz assigned to musical instruments as he put them to work in a scientific context. Music was both the means and the object of experimental investigation. Musical instruments in a scientific context, while certainly used as instruments of science, also maintained aspects of their musical selves. The changing design and increasing dominance of keyboard instruments in the musical world, the piano, harmonium, and organ especially, are thus central to this story.[4] And just as the musical instruments—both their design and techniques for playing them—were in flux, so too were the relationships between them and their users. This practice only increased as investigators began to examine the relationship between sound sensation and musical aesthetics in the mid-nineteenth century. This status of musical instruments in the laboratory, both as instruments of music and instruments of science, was central and necessary to the development of Helmholtz's physiological theory of tone sensation.

The second feature of Helmholtz's work that allowed him to reconcile a lawlike theory of sound sensation with a historicist understanding of musical aesthetics predated and extended beyond his laboratory: Helmholtz's own relationship with music and musical instruments. Music and musicianship had been a part of his life from a very early age. In an 1838 letter to his parents, the young Hermann Helmholtz declared that he always appreciated music more when he played it for himself.[5] In letters to his parents, to his two wives, and to friends and colleagues, Helmholtz wrote of his own experience of musical sound in a very personal and tender way. Helmholtz's initial encounter with music was the very intimate, personal exercise of playing music for himself.

So, this discussion of Helmholtz's work on sound sensation will begin with these origins: his initial relationship with musical instruments and with music itself.[6] Just as the performers and composers that followed Helmholtz's work so closely did not initially approach the piano as a scientific device, neither did Helmholtz. It was for him first an instrument of expression and enjoyment.

To these ends, this chapter mobilizes two larger historiographical themes. The first, the role of Helmholtz's sensory interaction with musical instruments, is an extension of current work in the history of science that promotes, indeed calls for, a "materialized epistemology" relating sensual with ideational knowing.[7] Recent scholars contend that rigid distinctions and dichotomies, between visual representations of nature and textual ones, neglect significant forms of knowledge making.[8] This chapter examines some of the ways music making was knowledge making in the nineteenth century. More specifically, it is an exploration of the ways Helmholtz's own musical training, his interaction with musical instruments, informed his scientific work. His relationship with music and musical instruments was multifaceted. In one sense he engaged them as a performer, physically and emotionally coaxing music out of his instruments. Additionally, Helmholtz's understanding of the aspects that contributed to the instrument's ability to make music extended beyond those of the performer and listener. These included not only an understanding of the mechanics of musical instruments but of the work of instrument makers. Helmholtz's physiological theory of sound sensation was both the product of, and constitutive of, how he heard and created sound.

The second theme this chapter explores is that of Helmholtz's classicism. A series of historical studies have highlighted the classicist nature of Helmholtz's scientific work. These works lend significant support to the categorization of Helmholtz's physiological, vision, and acoustic studies as classicist.[9] Similar practices and goals can be seen in his work on sound sensation. Helmholtz's musical aesthetics—his dislike of equal temperament, his favorite composers, his concertgoing practices—were classicist as well. This musical classicism was part of the larger neoclassical revival of the nineteenth century that informed Helmholtz's classicist science. Helmholtz's classicism, both aesthetic and scientific, in combination with his layered relationship with music and musical instruments, was critical to his reconciliation of universal physiological laws of tone sensation with his historically and culturally contingent aesthetics of music. Helmholtz's lifelong multilayered relationship with music—playing music for himself as well as attending concerts, his classicist aesthetic and listening practices, and his knowledge of instrument making—allowed him to fashion a sonic world around himself that made his theory of sound sensation possible.

In applying these historiographical concepts to Helmholtz's sound sensation work, this chapter is an examination analogous to the historical studies of Helmholtz's epistemology in relation to his physiological optics.[10] Most directly relevant is Timothy Lenoir's analysis of Helmholtz's empiricist understanding of vision, the epistemology of realism, and the construction of disparate elements of the sociocultural milieu into a new social and political order as interconnected through a common discourse.[11] This chapter explores such an intersection of ideas, specifically aesthetics and physiology, as it occurs in a single individual. Helmholtz was the embodied, psychophysical reconciliation of his multivalent participation in both the world of science and the world of music

For the sake of transparency it should be noted that Helmholtz would not have called himself a psychophysicist or his project psychophysical.[12] He would have seen Fechner's psychophysical experimentation in the 1860s and 1870s as studies of sense discrimination. But Helmholtz was not interested in issues of pitch discrimination. For him, a study of sound sensation was done through an examination of sound as an external, physical object. This chapter shows how Helmholtz also thought of sound in musical terms. Further, I argue that it was Helmholtz's particular musical tastes as well as his deeply personal interaction with musical instruments that allowed him to reconcile his conception of sound as physical object with his conception of sound as music by framing a sonic world for himself that consisted of very specific sounds. So psychophysics as I have framed it in this book—oriented toward aesthetics—occurred at this point of reconciliation.

Listening as a Musician

The epigraph at the beginning of this chapter is from Helmholtz's university years. Music had been part of Helmholtz's life from an early age. His parents saw music training as a form of both individual and social improvement.[13] He received piano lessons and even took a piano with him to the university. In response to a letter from Helmholtz in which he described his roommate's virtuosity on the piano, Helmholtz's father replied:

> I only hope your comrade is a stout-hearted, industrious lad; if he is, it will be great luck for you. His playing the piano so well is your best chance of improving yourself, and do not be so accommodating as to leave all the playing to him because he does it better than you, for it was under similar circumstances that I forgot all I ever learned: and, above all, don't let your taste for the solid inspiration of German and classical music be vitiated by the sparkle and dash of the new Italian extravagances—these are only a distraction, the other is an education. Be thankful for Cousin Bernuth's lessons, even if they are given somewhat crudely; behind these conventional social forms there lies in reality a deeper meaning, which is forgotten though it is still there, so that the form helps people to get on in society; to give them life so that they cease to be empty form and convention is the task of the individual.[14]

Helmholtz reassured his parents that he would not give up his music, that the only way for him to hear the music he liked was to play it himself, stating that "other people's expression and execution seldom satisfy me; I always care much more for music when I am playing it myself."[15] While at the university he devoted any spare time during the day to music, about an hour a day during the week, more on the weekends. Alone Helmholtz would play Mozart and Beethoven sonatas but he would also sight-read through new sheet music together with his roommate.[16] He would go to great lengths to hunt down this new sheet music; Mozart and Beethoven sonatas were always particularly prized finds.[17]

Helmholtz's familiarity with, indeed love of, playing the piano informed the way he heard and understood music. As a diligent player he would have known well the physical manipulations required of his hands and body to generate a musical sound acceptable to his tastes—both the correct notes and the desired coloration of tones. Helmholtz's constant playing likely also motivated his intimate knowledge of the mechanics of the piano and the role played by each of the various components of the hammer action system in generating sound.[18] So his understanding of how the piano generated music was personal, corporeal, and sensory as well as intellectual.

Helmholtz was similarly well acquainted with other musical instruments. Many of his acoustic experiments made use of the violin, harmonium, and zither. His 1860 article, "On the Motion of the Strings of a Violin," was devoted to refuting Young's claim that the strings of the violin vibrated in a complex and irregular way. He complained of his struggle to get pure tones from inferior violins. But with a Guadagnini violin, "and after some practice, [he] got the curve completely quiescent as long as the bow moved in one direction, the sound being very pure and free from scratching."[19]

Later, in *Tonempfindungen,* Helmholtz was able to show that the jumps and displacements of the otherwise regular form were due to the scratching of the bow, especially during the beginning of the sounded tone when the bow first made contact with the violin string. He claimed that these discontinuities in the vibration curve "betray every little stumble of the bow."[20] For Helmholtz the ability to coax proper tones—not simply sound, but musical sound—from musical instruments was a requirement for acoustic studies. Indeed, in his 1869 address, "The Aim and Progress of Physical Science," he noted that the experimental investigator must have sharpened senses as well as hands trained to do the work of, among others, the violin player.[21] According to Helmholtz, the experimental scientist required the refined hearing and touch of the music performer.

In an 1865 letter to his second wife, written from Paris, Helmholtz described a concert he attended and was particularly impressed by the orchestra:

> One hears better choral singing in Germany, but the perfection of the orchestra is unique of its kind. The oboes in Haydn's Symphony sounded like a gentle zephyr; everything was in perfect tune, including the high opening chords of the Mendelssohn Overture, that are repeated at the end, and generally sound out of tune. The *Prometheus* was the most enchanting melody, with the horns predominating. This concert, after the Venus de Milo, was the second thing of purest beauty that life can give.[22]

This quote illustrates Helmholtz's great love for music. It also illustrates his highly skilled musical ear. Helmholtz was confident in his ability to judge the intonation of the opening, multipart chords sounded by several different instruments. Further, this quote reveals the extent of his familiarity with the classical musical repertoire; he knew the Haydn piece so well that he could anticipate the difficult passages.

In an 1860 article on Pythagorean intonation Helmholtz carefully described the slight adjustments made to the twelve-tone scale that resulted in less-than-pure intervals. He admitted that the theoretical exactitude (a Pythagorean third interval ratio of 64:81 versus a pure third interval ratio of 64:80, for example) was negligible to many and could only be distinguished by those individuals with a "skilled musical ear," clearly including himself among them.[23] Again, the experimental skills of the laboratory are seen overlapping with Helmholtz's musical training. He was—especially for an individual initially trained on a keyboard instrument—highly sensitive to intonation, tone color, and the nuances of orchestration. And given that Helmholtz repeatedly and explicitly encouraged experimentalists to develop the refined hearing and touch of the music performer, he clearly believed that musicians maintained skills distinct from those of the average experimental scientist. Those who listened in the laboratory with the training of a musician heard and generated sound differently, better. When Helmholtz listened to sound, he listened as a musician. He had an instrumentalist's ear.

Listening as a Classicist

As discussed previously, significant historical work lends support to the categorization of Helmholtz's science as classicist, showing that idealization and ideal types were central to his scientific work and that he maintained a larger classicist goal of revealing universal truths. Similar scientific practices and goals can be seen in his work on sound sensation. Further, Helmholtz's musical tastes—the composers he liked, the equal temperament he disliked—were classicist as well. Specific tuning systems and specific musical works generated specific tones and harmonies—for example, a Haydn string quartet performed on just-tuned instruments would have had a specific sound. As was the growing trend by the mid-nineteenth century within certain social strata, such a performance would be listened to and appreciated in a specific way, as part of a classicist enterprise of searching for ideal forms within the musical work. Helmholtz listened to music in such a way, as a classicist. Not only was his scientific practice classicist, as other historians have asserted, but his aesthetics were classicist as well. His classicist musical aesthetics, in turn, informed his science; if he sought only certain tones and harmonies and only listened to them in a specific manner, his understanding of sound sensation would be similarly circumscribed.

Recall from the previous chapter that the mid-nineteenth-century music critics and music theorists had begun to cultivate and promote a new practice of listening to music. The writings of such music critics as Adolf Bernhard Marx and Eduard Hanslick advanced theories of musical aesthetics that called for a new kind of listening. Both critics emphasized musical form as the essence and determiner of content. Marx's bald idealism was the basis of an analytical system that allowed him to validate and elevate the formal procedures of composition and their use of the classical style, particularly in the music of Beethoven.[24] Hanslick's formalism, as presented in his 1854 treatise, *Vom Musikalische-Schönen*, was rooted in the belief that beauty had no object or aim beyond itself. Thus music had no goal of giving the listener pleasure, nor was music in need of an extramusical subject introduced through, for example, the use of words. The sonically moving forms were both the essence of music and functioned to express musical ideas.[25] The musical idea was an object of intrinsic beauty independent of its reproduction by a performer. For Hanslick, it was the forms that the listener actively contemplated, delighting in the musical structures.[26]

Concertgoing practices shifted to accommodate this intellectual approach. Audiences would sit quietly, carefully listening not just to enjoy music but to understand it.[27] So, though audiences may have been listening to what would be termed romantic music their listening practices were otherwise. Their experience of the music was not a rapturous, emotional one but rather a thoughtful exercise in which the music-theoretical aspects of the piece were carefully contemplated.

Historian William Weber has described the early nineteenth-century music audience as divided into two "taste publics."[28] The upper middle class and the aristocracy tended to favor flashy virtuosic pieces and popular operas in which the performance of the piece, rather than the score, was the aesthetic arbiter. The liberal professionals—the *Bildungsbürgertum*—saw such pieces and performances as tasteless and were highly critical of the perceived commercialism of popular opera and the cult of the virtuoso. (Recall the warnings of Helmholtz's father against the young Helmholtz getting drawn in by his roommate's tasteless flourishes on the piano.) The Bildungsbürgertum instead maintained a "highly ascetic attitude" toward music in general and their listening practices reflected this.[29] They preferred to listen to the works of Haydn, Mozart, Beethoven, and Schubert, and when they did it was an intellectual exercise rather than an emotional experience.[30]

This Bildungsbürger taste public was also central to the consolidation of a German national culture in music, especially if one includes as well the choral music scene, which was composed largely of the lower middle class but had direct links with the classical music scene. The classical repertory movement, the Bach revival, the celebration of folk song, all contributed to the view that German music was superior to all others precisely because it was universal and culturally transcendent.[31] Recent scholarship has further argued that the flourishing of autonomous art in the nineteenth century was in part due to autonomous art's dual role as both purely aesthetic object as well as medium through which to pursue certain social projects.[32]

Helmholtz was part of the Bildungsbürger taste public. He would have approached a musical performance as an object to be carefully contemplated according to the rules of formalism. This becomes, then, a discussion of how Helmholtz heard. And what he heard. His musical aesthetics and his practice of science can be understood to be preoccupied with issues of form and universal laws, perhaps echoing the formalist listening goals promoted by several music theorists and music critics at the time (Marx and Hanslick especially). For Helmholtz, music was a valid avenue through which to approach and understand the sensation of sound; music and sound were treated as interchangeable scientific, investigative objects. Both had materiality and form. Both were, according to contemporary understandings of form, historicist.

I want to refer briefly here to Norton Wise's work on the network of bourgeois Berlin—the personal relations and institutional affiliations connecting and supporting the young men who formed the Berlin Physical Society in the early 1840s.[33] Connecting Helmholtz's frog-drawing experiments to his conservation principle and its embodiment in his instruments, Wise suggests he had a twofold goal. First, according to Wise, Helmholtz hoped to mobilize the instrumental embodiment of the connection between frog-drawing experiments and the conservation principle to illuminate the elusive concept of *Energie*. Second, Wise argues that Helmholtz sought to capture the fundamental *form* of the temporal processes of the muscle. This ideal form for Helmholtz would be expressed in a curve of Energie, as though drawn by the frog muscle.[34]

To this end Wise documents Helmholtz's early engagement with the theory and practice of drawing, from his Gymnasium drawing classes to his time teaching anatomy at the Academy of Art. Wise finds evidence that Helmholtz borrowed many ideas from his father, Ferdinand Helmholtz. Ferdinand Helmholtz's Gymnasium curriculum was structured around a highly idealist, grand humanist vision that sought the intuitively and completely expressed idea or form (which he identified with beauty) of an object through the vigilant documentation of its details. For both Helmholtz and his father, this intuitively expressed form was possible through proper drawing.[35] Wise suggests that Helmholtz can be thought of as teaching the frog muscles to draw their own changing states, to express through their form the idea Helmholtz held of their states of Energie and of their beauty (according to his own aesthetic guidelines). In his laboratory work, Helmholtz relied on his extensive training in drawing, his belief that properly detailed formal aspects could express the form of the idea, and his corresponding belief that the proper analysis of the formal aspects of a curve could reveal the idea within. So where Wise finds Helmholtz's engagement with the theory and practice of drawing in his early life manifested in his goals of capturing the fundamental and ideal form of the temporal processes expressed by the frog muscle, I similarly find Helmholtz engaging with concepts of form in his listening practices as well as in his very definition of sound.

Admittedly, at first blush, Helmholtz's personal attitude toward music might appear romantic, especially his belief in the primacy of music above all other forms of art.[36] Helmholtz did appreciate the emotive aspects of music. He explained that it was a medium that permitted an emotional interaction between performer and audience such that the audience was moved by the "outflow of the artist's own emotions."[37] In a letter to his first wife, Helmholtz expressed his feelings for her in terms of music, addressing her as his nightingale with the sweet and low song of *Alceste*, with whom to sweetly stroll through beautiful Beethoven-like gardens.[38] He declared in 1865 to his colleague Carl Ludwig that Beethoven was the "mightiest and most moving of all composers" and that he rarely played anything else.[39] Helmholtz had effusive praise for Beethoven's *Coriolan Overture*, with its unrest and battle, ultimately resolved to a pair of the most melancholy tones, saying that "there could be no greater masterpiece."[40]

This language appears to be at odds with the contemplative and formalist goals of aesthetic listening promoted by Hanslick. But for Helmholtz music was not mysterious or unknowable. And though he appreciated the emotional power of music, Helmholtz emphasized the formal aspects of harmony and melody. In his 1865 letter to Ludwig, Helmholtz elaborated on his defense of Beethoven's genius as a composer, stating:

> I too find [Beethoven] the mightiest and most moving of all composers, and I myself play hardly anything but Beethoven, when I do play. Had I been speaking about the vehicle of musical emotion, I should certainly have placed him above all others. I was, however, talking exclusively of melody, and the fine artistic beauty of the flow of harmony, and there I do hold Mozart to be the first, even if he does not affect us too powerfully. Speaking generally, as one grows older, and bears more scars within one's breast, one ceases to feel that emotion is really the greatest thing in art.[41]

Helmholtz clearly appreciated not only the emotive aspects but also the *intellectual* aspects of harmony and melody. Mozart was to be respected for the perfection of his harmonic and melodic structures. For Helmholtz, fine harmony and melody were aesthetic qualities independent of the emotion they did or did not generate in listeners. The formalist aesthetics of harmony and melody were just as, if not more, significant in the art of music than the emotions they elicited.

I have claimed that Helmholtz's instrumentalist's ear, among other things, shaped what he heard and shaped what he thought. So did his classicist musical tastes. A close reading of two decades of Helmholtz's letters to his first and second wives, while by no means painting a complete portrait, provides some further insight into Helmholtz's concertgoing classicist tastes. His correspondence reveals what musical works he sought out and what features about them he appreciated and so, by extension, shows that Helmholtz did indeed listen to music in a formal, classicist manner.

Beyond his home performances, which likely occurred almost daily, Helmholtz clearly enjoyed public musical performances as well. He would —if one includes church concerts, choral society performances, and rehearsals by professional ensembles—attend several in a given week. He frequented opera as much as orchestral music, and oratorios and incidental music most of all, though this may have been because choral music was performed more often.[42] Although he attended three operas that could be classified as Italo-French (Rossini's *Il barbiere di Siviglia*, Butera's *Atala*, and Auber's *Le Maçon*), nearly every piece Helmholtz referred to in his letters was by a composer from the German-speaking world: Beethoven, Mozart, and Mendelssohn most of all. With the notable exception of the works of Liszt and Mendelssohn, none of the pieces Helmholtz discussed were by living composers. His concertgoing practices coincided with the general nineteenth-century push to restore or reconstitute the so-called classical repertoire (from the previous chapter, recall the local effort by Marx to train the Berlin listening public through repeated performances of Beethoven). In this period the presence of earlier music became overwhelming, a fundamentally new feature of European music culture, and a product of romanticism's push to celebrate the canons of certain composers as exemplary of particular musical genres.[43]

Helmholtz was part of this effort to reify certain composers of Germany and Austria as exemplary, as classical. Recall that Helmholtz described the oboes of the Haydn symphony as "a gentle zephyr," the melody of Beethoven's *Prometheus* as "the most enchanting."[44] He regarded an otherwise poor performance as redeemed by an "extremely magnificent excerpt" of Mendelssohn's forgotten opera *Die Lorelei*.[45] And as noted previously, Helmholtz found Mozart to be the greatest composer of harmony and melody.[46] His great works contained, according to Helmholtz, "profoundly musical internal harmonies" such that his ears heard only "musical figures."[47] The beauty of a musical performance, for him, was located in the composition's melodic and harmonic structures, the musical figures. Helmholtz did not listen in a state of rapturous emotion but rather one of careful contemplation, attentively seeking and analyzing formal features. Hanslick would have approved.

Like other formalist listeners, while Helmholtz celebrated good composition over good performance, he was critical of virtuosic concerts as well as the Italo-French operatic style. He was dismayed that while his university roommate was very talented at the piano, he insisted on only playing florid, modern Italian music.[48] Twenty years later, in 1863—the same year Liszt performed before the Pope—Helmholtz remained unimpressed by virtuosity and described a performance of Liszt's *Symphonic Preludes* to his wife as "effective and extraordinary enough, but hardly beautiful."[49]

To describe Helmholtz's musical tastes as classicist is not merely to say that he appreciated music by composers generally termed classical, though many of them were. Helmholtz's listening practices were those of the mid-nineteenth-century, liberal professional, classically trained "taste public." His letters to his wives and others indicate that he appreciated good composition over good execution. For Helmholtz, fine harmony and melody were aesthetic qualities independent of the emotion they generated in listeners—he made a distinction between the genius of a composer's harmonic and melodic choices and their ability to generate emotional responses in listeners. Helmholtz clearly appreciated not just the emotive aspects but also the intellectual aspects of harmony and melody. He listened to music actively and thoughtfully, as Hanslick instructed—a classical enterprise indeed.

Another component of Helmholtz's classicist musical taste was his strong preference for just intonation. During this period the European musical world was struggling to transition to a more performance-friendly equal temperament tuning system (see chapter 2 for a more detailed discussion). Helmholtz was a highly vocal critic of all forms of tempered intonation, equal temperament and Pythagorean intonation in particular. The equally tempered thirds, for example, were located somewhere between pure thirds and Pythagorean thirds, and thus, according to Helmholtz, "correspond to no possible modulation, no tone of the chromatic scale, no dissonance that could possibly be introduced by the progression of the melody; they simply sound out of tune and wrong."[50] He continued, explaining that in slow passages, higher-pitched thirds sound particularly out of tune with the bass and "hence sounding as if they were played on some other instrument which was dreadfully out of tune."[51] Helmholtz took it on himself to campaign for the cause of just intonation, in which pure thirds and pure fifths were maintained (the scale was *un*tempered).[52] This was, however, a doomed project, more the product of nostalgic yearning, because most pianos were tuned to equal temperament by the 1870s.

In 1860 Helmholtz had proposed a design for a just-tuned harmonium.[53] Later, because of its importance "for the solution of many theoretical questions to be able to make experiments on tones which really form with each other the natural intervals required by theory, to prevent the ear from being deceived by the imperfections of the equal temperament," Helmholtz commissioned the construction of such an instrument.[54] Its two sets of brass reeds (metal tongues that vibrated as air flowed over them) were tuned to include fifteen major chords and fifteen minor chords. As required by just intonation, the thirds were all perfectly pure. The purity of the fifths had to be sacrificed in order to modulate between all thirty scales on the two sets of reeds and were thus slightly flat, though only by one-eighth of the interval by which equal tempered fifths were flatter than pure fifths.[55] Compared to an equal-tempered harmonium, Helmholtz claimed the difference of musical effect to be "very remarkable." The just-tempered chords of Helmholtz's instrument "possess a full and as it were saturated harmoniousness; they flow on, with a full stream, calm and smooth, without tremor or beat."[56] Equal-tempered or Pythagorean-tuned chords, however, "sound beside [just-tempered chords] rough, dull, trembling, restless."[57] This was apparent even to those not musically trained. Though Helmholtz admitted that equal temperament was extremely advantageous to instrumental musicians (eliminating the requirement that they retune their instruments every time they needed to play in a different key), and that many modern musicians had heard nothing but equal-tempered music, he insisted, "it must not be imagined that the difference between tempered and just intonation is a mere mathematical subtlety without any practical value."[58]

Helmholtz noted that the beats that arose from equal-tempered thirds were clearly audible and particularly disturbing around middle pitches, on instruments with harsher tone quality, and during a slow musical passage. Thus, rapid passages performed with soft quality and moderate volume of tone revealed little of the "evils of tempered intonation."[59] The use of these flashy, virtuosic passages to hide such evils likely fueled Helmholtz's loathing of them. Further, their increasing prevalence did not bode well for the future of music. Helmholtz observed that nearly all contemporary instrumental music was "designed for rapid movement" and raised the question of whether instrumental music would, in fact, be forced by tempered intonation to renounce the full harmoniousness of slow chords for rapid passages by tempered intonation.[60]

The effects of tempered intonation varied by instrument; on the piano, for example, they were less marked. But they were still noticeable. Helmholtz claimed that when he went from his just-tuned harmonium to a tempered piano, especially for slow-moving chord passages, "every note of the latter sounds false and disturbing."[61] Further, he continued, for the music performed by the very best instrumentalists (on bowed and wind instruments where the intonation can be controlled by the players themselves) it was impossible to detect "any false consonances."[62] Performers naturally played in just intonation. Helmholtz claimed "the only assignable reason for these results is that practiced violinists with a delicate sense of harmony, know how to stop the tones they want to hear, and hence do not submit to the rules of an imperfect school."[63] Further, amateurs that learned to sing or play their instrument independent of tempered accompaniment or instruction were also inclined to perform in just intonation.

Tempered intonation not only forced performers against their nature, Helmholtz lamented, it also threatened the very art of music; it threatened the ideal form of the pure harmonic intervals. With the imperfection of consonant chords, new and unusual harmonies, inversions, and tonal modulations were required for expression within particular keys. Even these inversions and modulations in an equal-tempered system were incomparable to the expressiveness of the equivalent passages in a just-tuned system.[64] The dissonant chords and "continual bold modulational leaps" of modern music threatened the "feeling for tonality," presenting "unpleasant symptoms for the further development of art."[65] So, the introduction of equal temperament was forcing composers to either compose fast, tacky pieces or to develop strange harmonies in order to establish the "moods" formerly achieved, in just intonation, by simply composing in specific keys. The great composers Mozart and Beethoven, Helmholtz continued, reigned prior to the extensive use of equal temperament. And they were, respectively, responsible for the "sweetest possible harmoniousness" and music effects "which none had hitherto attempted."[66] A coincidence? Helmholtz thought not. For him, tempered intonation and the convenience it offered musical instrument design, threatened "the natural requirements of the ear, and to destroy once more the principle upon which modern musical art is founded."[67]

This quote must be unpacked. According to Helmholtz, the ear had natural requirements that coincided with the foundations of modern music. This modern music was, for Helmholtz, the reign of Mozart and Beethoven, prior to the introduction of equal temperament. The consequences of Helmholtz's classicist musical aesthetics for his scientific study of sound sensation are now apparent. Helmholtz limited the sounds he heard according to his preferences for classical compositions—those that lent themselves to formal analysis—as well as for just intonation. Accordingly, Helmholtz asserted that sound sensation accommodated these sounds and these sounds only. The newfangled virtuosic performances, harmonies, and temperaments against which Helmholtz railed, did not meet the natural requirements of the ear and the sensation of sound.

Listening with an Understanding of Musical Instrument Design

Not only did Helmholtz frame his sonic world according to his instrumentalist ear and classicist musical aesthetic, he also incorporated his knowledge of musical instrument construction. Helmholtz maintained a strong technical and mechanical understanding of musical instruments, especially their manufacturing. He worked closely with musical instrument makers, not only to improve his technical understanding but also to commission such custom instruments as the previously mentioned just-tuned harmonium. In some cases the workshops specialized in both musical and scientific instruments.[68] Helmholtz's engagement with instrument makers and the technical side of instrument production—and Helmholtz stated this explicitly—reinforced his theories about the nature of sound. In commissioning new instruments as well as contributing to innovations in instrument design, Helmholtz moved beyond his generative abilities as a musical performer in which he would be limited by traditional instruments. In his involvement with the development of new musical instruments, Helmholtz was able to surround himself with instruments that generated sounds reinforcing his theories of consonance and dissonance as well as sound sensation.

Helmholtz distinguished between keyboard instruments and others. The former were highly constrained by the mechanics of their design. Every single note on a keyboard instrument was fixed. But the quality of the sound generated on bowed instruments, for example, was limited only by the relative tuning of their (four) open strings.[69] In his lecture before the Glasgow Philosophical Society, Helmholtz had noted the difference between the bowing and the plucking or striking of strings. Bowing did not excite harmonics at the point of contact of the bow on the string as pianos and harpsichords did.[70] The sounding overtones on the plucked string resulted in a tone that was not only less pure but more likely to be subject to interference beats. The purity of a tone sounded on a bowed string could be controlled. By the third edition (1870) of *Tonempfindungen*, Helmholtz had come to understand that this freedom from temperament constraints inherent in bowed instruments as well as the variety of tone quality permitted by the bow, allowed for more artistic freedom for the player: "The art of bowing is evidently the most important condition of all. How delicately this must be cultivated to obtain certainty in producing a very perfect quality of tone and its different varieties."[71] Myles Jackson has argued that Helmholtz meant here to illustrate how proper musical bowing, because it was a cultivated skill, could not be simply reduced to mechanics.[72] String instrument performers, in their control over tone pitch and color, brought their individual touch to their instruments.

Keyboard instrument performers, however, could not express their individuality to the same degree. In the case of keyboard instruments Helmholtz instead emphasized the role of the instrument *maker* in sound quality. He had taken great interest in the work of keyboard instrument makers. In the fall of 1865, Helmholtz visited the workshop of the organ builder Cavaillé-Coll. He was given a tour of the workshop and was taken to visit the Church of Saint-Sulpice, which housed an organ built by Cavaillé-Coll, then the largest organ in Europe. Helmholtz described the experience as highly interesting, and he commended the artisan for his intelligence and originality and for raising himself to the level of master. He also took the time to proselytize to the builder about the virtues of just intonation.[73] They then went to the house of the harmonium maker Mustel to see his latest invention, a tuning-fork piano with sustained tones. Helmholtz wrote to his (second) wife: "This confirmed my theoretical assumptions, and produced no special effect, which fact, however, is of some importance for my theory. The advance in the construction of the harmonium was very striking; it was like a very perfect and easily responding piano, with every kind of contrivance in the mechanism, for bringing out the treble parts."[74] In 1885, an American pianist and piano pedagogue asked Helmholtz the extent of the pianist's control over the quality of the instrument's sound, to which Helmholtz replied: "As far as I know, on the newer mechanisms the rate of speed with which the hammer flies against the string, i.e., the force of the blow from the key, is the only way to modify tone."[75]

For strings excited by striking, such as those on a piano, Helmholtz explained in *Tonempfindungen*, the force of the upper partials (overtones) was dependent on the nature of the stroke, where the string was struck, and the density, rigidity, and elasticity of the string.[76] A sharper, shorter stroke—striking the string with a sharp-edge metallic hammer that instantly rebounds, for example—resulted in a longer series of upper partial tones that could have intensities that surpassed even the fundamental tone. A softer, more elastic hammer would result in fewer. Helmholtz encouraged the reader to open the lid of a piano and compare plucking a string with his or her finger, striking it with a metallic edge, and striking it with the piano hammer.[77] Again, the skills of the musician and the skills of the acoustician were complementary. Sound could be understood and discussed in musical terms.

Helmholtz continued, explaining that heavier hammers with thicker felt were used on the lower strings of the piano and lighter hammers with thinner felt were used on the higher in order to balance the upper partial tones that would sound when the strings were struck. It was therefore, according to Helmholtz, the piano makers that controlled the quality of the instruments' tone.[78] He elaborated: "Clearly the makers of these instruments have here been led by practice to discover certain relations of the elasticity of the hammer to the best tones of the string. The make of the hammer has an immense influence on the quality of tone."[79]

Helmholtz believed that the quality of musical tones, deeper tones especially, could be improved if the higher of the upper partials that inevitably sounded along with a struck note on the piano were eliminated. He described a new grand piano designed by the Steinway piano makers of New York that he found "remarkable for the evenness of its quality of tone," explaining that

> the damping resulting from the duration of the strike falls, in the deeper notes, on the ninth or tenth partials, whereas in the higher notes, the fourth and fifth partials were scarcely to be got out with the hammer, although they were distinctly audible when the string was plucked by the nail. On the other hand upon an older and much used grand piano, which originally shewed the principal damping in the neighborhood of the seventh to the fifth partial for middle and low notes, the ninth to the thirteenth partials are now strongly developed. This is probably due to a hardening of the hammers, and certainly can only be prejudicial to the quality of tone.[80]

Though Helmholtz and Theodore Steinway maintained a correspondence over several issues of piano design, the true extent of his relationship with Steinway is difficult to determine.[81] The Steinways certainly would have encouraged the impression of a close relationship if only for the prestige associated with Helmholtz's endorsement of their pianos. They were no doubt pleased in 1871 when Helmholtz chose, after examining several different pianos, to use a Steinway grand for his experiments and lectures in Berlin. That year he wrote to Theodore Steinway to thank him for the instrument and complimented the "prolonged vibration of its tones, by which the instrument becomes somewhat organ-like" as well as the "lightness and delicacy of touch."[82]

The following year, Theodore Steinway applied for a patent titled "Improvement in the Duplex Agraffe Scales for the Piano-Forte," a stringing innovation he referred to as the "duplex scale." Prior to this innovation a "dead" or "waste" end of string between the bridge pin and the hitch pin secured the "live" or vibrating length of a piano string. Piano makers attempted to limit the dissonant sounds generated by these vibrating end strings by damping them with felt. Theodore Steinway's innovation was to remove this felt and adjust the length of these unused end strings so they could resonate sympathetically with the live string. He claimed this resulted in a fuller sound in the upper register.[83] Richard K. Lieberman has argued that Theodore Steinway developed the duplex scale on the basis of his reading of Helmholtz.[84] There is some indication that Helmholtz actually suggested such design innovations to Steinway, though nothing concrete.[85]

In 1885 the Steinways gave Helmholtz another piano, a style B drawing-room grand. The model had been introduced the year before and would have included the duplex scale innovation. Of this gift Helmholtz wrote to William Steinway: "With such a perfect instrument as yours placed before me, I must modify many of my former expressed views regarding pianos" and "I have repeatedly and carefully studied the effects of the Duplex Scale . . . and find the improvement most surprising and favorable, especially in the upper notes, for splendid as my Grand Piano was before, the Duplex Scale has rendered its tone even more liquid, singing and harmonious."[86] The relationship between Helmholtz and the Steinway family (at least as it was presented by the Steinways, who, to put it cynically, had much to gain commercially from touting Helmholtz's support) nicely illustrates Helmholtz's deep and intricate knowledge of the mechanics of piano design.

Steinway pianos were distinctive for their Victorian aesthetic: a heavier, muted sound.[87] Helmholtz clearly appreciated this aesthetic, especially after Theodore Steinway's duplex scale innovation increased the coincidence of harmonic overtones. Helmholtz had used a Kaim and Günther piano in his early work on tone sensation but, as stated above, he began employing Steinway pianos for experiments and lecture demonstrations in 1871.[88] His letters to the Steinways indicate that he found their introduction of the duplex scale to result in a more beautiful sound. The design innovations of the Steinway pianos and the late nineteenth-century Steinway sound would have not only reinforced Helmholtz's musical aesthetics but also contributed to his theory of harmonic consonance (to be discussed shortly) that was rooted in the coincidence of harmonic overtones.

In the introduction to *Tonempfindungen*, Helmholtz had carefully noted that the investigations described would be impossible for an individual to understand without personal observation of the phenomena described. He continued: "Fortunately with the assistance of common musical instruments it is easy for anyone to become acquainted. . . . Personal observation is better than the exactest description, especially when, as here, the subject of investigation is an analysis of sensations themselves."[89] Certainly Helmholtz's personal, sensory interaction with musical instruments—both as a performer and in his engagement with the technical side of instrument production—contributed to and reinforced his understanding of tone sensation. Indeed, his relationship with musical instrument makers allowed him to surround himself with a number of instruments (his commissioned just-tuned harmonium, his Steinway style B grand) that generated very specific musical sounds. Regardless of whether he employed these instruments for personal music performance or sound sensation experiments, their presence almost certainly reinforced his belief that classical harmonies and just intonation met the "natural requirements of the ear." Helmholtz's mechanical understanding of musical instrument design shaped what he heard. And what he heard, in turn, informed the development of his theory of tone sensation.

Playing Music for Oneself

When he played the piano for himself, both generating and receiving the musical sounds of the piece, Helmholtz guaranteed himself certain sound sensations. Because he framed his sonic world in a specific way, he correspondingly framed his understanding of sound sensation in a specific way. To summarize: to successfully convey both the localized effects and the deep structural "idea" within, say, a Beethoven sonata in a musical performance (if only for himself), Helmholtz would have depended on two features of his piano. First, the idea conveyed, reinforced by the key of the piece, carefully chosen for its specific mood, would only be heard as such on a piano *not* tuned to equal temperament as was very likely the case. Further, if Helmholtz played the piece for himself on, say, his Steinway style B grand, complete with its new duplex scale innovation that reinforced consonant partial tones, the harmonious effects of Beethoven's genius would have been further amplified. A circularity becomes apparent: between Helmholtz's practiced piano touch voicing the piece in such a way as to draw out the significant harmonic structures, the just intonation, and the reinforced consonant partial tones of the duplex scale feature, he would have both generated exactly the sounds he was listening for and normalized them as natural.

Let us now widen the aperture and consider whether, when he wasn't the one generating the music, if Helmholtz could still hear the natural sounds he was listening for. In his discussion of compound tones and the development of the third interval, Helmholtz admitted that, both when listening to melodies in which pure thirds were employed and when observing isolated, sounded intervals, he could not be certain that he was experiencing a natural third (a frequency ratio of 4:5). He was certain that the natural third was "calmer" (*ruhigeren*) than that of the equal temperament (which he described as "sharper") or Pythagorean tuning (sharper still).[90] And while modern musicians, because they were accustomed to equal temperament, preferred equal-tempered thirds, Helmholtz was convinced that performers of the first rank remained committed to natural thirds, not just in harmony but melody as well.[91] He referred to the work of Delezenne as having already established that top musicians played in just intonation.[92]

Helmholtz presented the violinist Joseph Joachim as his case in point. Between the initial publication of *Tonempfindungen* and the third edition, Helmholtz repeated Delezenne's observations, testing on Joachim.[93] He asked the violinist to tune his instrument's strings to Helmholtz's just-tuned harmonium. He then asked Joachim to play a G-major scale. The violinist played both the third and the sixth intervals as pure relative to the starting note, *not* equally tempered. Helmholtz explained that such masters "will readily convince themselves by the testimony of their ears, that the facts here adduced are correct, and perceive that far from being useless mathematical speculations, they are practical questions of very great importance."[94]

When Helmholtz asked singers—who would have been used to being accompanied by an equal-tempered piano—to sing a melody unaccompanied, he found they sang pure third and sixth intervals.[95] Even when trained to adjust to their equal-tempered accompaniments, singers would, when given the opportunity, naturally sing pure intervals. Musicians who performed on instruments without fixed tones, vocalists and string instrumentalists especially, were guided only by their ears.[96] Helmholtz continued, arguing that no doubt could remain that "the intervals which have been theoretically determined . . . are really natural for uncorrupted ears; that moreover the deviations of tempered intonation are really observed and disliked by uncorrupted ears; and lastly that, notwithstanding the delicate distinctions in particular intervals, correct singing by natural intervals is much easier than singing in tempered intonation."[97]

Musicologist Ross Duffin reasonably points out that Joachim may have simply been obliging Helmholtz.[98] Then again Duffin also suggests that because Joachim performed so frequently with quartets, where the harmonizing intervals between the different instrumental lines would have been pure, it was likely that his solo performances similarly tended toward pure intervals.[99] Indeed, Joachim employed a tuning in his solo performances distinct enough that George Bernard Shaw later termed it the "Joachim mode."[100]

Joachim was born near what is now Bratislava, though he moved while quite young to Pest. He was sent to study at the Vienna Conservatory to further nurture his already significant talent on the violin, and then, at twelve, he began studying with Mendelssohn in Leipzig. On Mendelssohn's death in 1847, Joachim joined Franz Liszt in Weimar, one of his early disciples of a new musical style opposed to the conservative Leipzig style. In 1849, the Hungarian violinist Eduard Reményi introduced Joachim to his accompanist, Johannes Brahms.

Joachim moved to Hanover and, in 1853, met Clara Schumann. Later that year he introduced to Brahms to the Schumanns who were immediately impressed with the young man's compositional abilities. Though Brahms had yet to publish any works, Robert Schumann devoted an 1853 article for the *Neue Zeitschrift für Musik*, titled "Neue Bahnen" ("New Paths"), to praising Brahms's genius. Robert Schumann would also be instrumental in encouraging Hermann Härtel to publish Brahms's compositions. Joachim, the Schumanns, and Brahms were the core of the Leipzig-based group of ascetic musicians that would eventually clash with the Weimar-based New German School (*Neudeutsche Schule*).

Figure 3.1
Joseph Joachim and Clara Schumann, by Adolph Friedrich Erdmann von Menzel.

Joachim and Clara Schumann would tour together extensively and represented a new ascetic style of music interpretation (figure 3.1). Individual glory through virtuoso technique was subordinated to the goals of the composer. Joachim especially was known for an austere repertoire. He stood, as Shaw described, "inflexibly by the classics," and concentrated on Bach sonatas (he performed the Bach D-minor Chaconne more than any other piece) and the violin sonatas of Beethoven, Mendelssohn, and Brahms.[101] Helmholtz likely would have enjoyed Joachim's concerts.

Musicologist Beatrix Borchard describes Joachim's playing (recordings from 1903 remain) as revealing a subtle command of rubato, a spare use of vibrato, and long-arched phrasing.[102] Hanslick identifies the most outstanding characteristic of Joachim's playing as his "modest, unadorned greatness" of sound, an "unbending, Roman earnestness."[103] Further testifying to Joachim's ascetic approach, he describes an 1861 Vienna concert as reflecting "truly enormous accomplishments of a brilliant but always subordinate technique" and that Joachim lacked the "vanity common to the virtuoso."[104]

By the mid-1850s, Joachim had completely broken with Liszt—with both his showy performance style and the aesthetics of what had become, in collaboration with Richard Wagner and Hector Berlioz, the "Music of the Future" (as it was then popularly called, later to be termed the New German School). Joachim described his disillusionment to Clara Schumann, admitting that "a more vulgar misuse of sacred forms, a more repulsive coquetting with the noblest feelings for the sake of effect, had never been attempted."[105] He continued, claiming that he could never again see Liszt for fear of insulting him to his face. Joachim actively eschewed all showmanship, as Hanslick noted when he effused over the 1861 Vienna concert: "For all his technique, Joachim is so identified with the musical ideal that he may be said to have penetrated beyond the utmost in virtuosity—to the utmost in musicianship. His playing is large, noble, and free. Not even the slightest mordent has the flavour of virtuosity; anything suggestive of vanity or applause-seeking has been eliminated."[106]

The subtext of Hanslick's accolades was an implicit denigration of virtuosic showmanship. Like Robert Schumann and Mendelssohn of the previous generation, such artists as Brahms and Joachim joined forces with Hanslick against both the performance and compositional aesthetic of the Music of the Future. In his 1859 inaugural address at the Tonkünstversammlung (what would become the Allgemeiner Deutscher Musikverein), Franz Brendel described the movement led by Wagner, Liszt, and Berlioz as representatives of the entirety of post-Beethoven musical development and therefore deserving of the title "New German School." Joachim and Brahms took great offense to Brendel's refusal to acknowledge the formalist aesthetic of performance and composition that developed mostly in Leipzig (the music of Mendelssohn, Schumann, Brahms, and Joachim and the music criticism of Hanslick). They circulated a "letter of refusal" among sympathizers for signatures, critiquing Brendel's characterization that the controversy had already been fought out and settled in favor of the so-called New German School, whose principles they deplored as "contrary to the most fundamental essence of music."[107] Subsequent embarrassment over the leak and publication of the document before it contained more than four signatures convinced Brahms, for one, to avoid such public battles in the future.[108]

Hanslick, however, stoked such public clashes. It was, of course, through his music criticism that Hanslick exercised his great power in the music scene of Vienna and beyond. The works he chose to review—mostly new works and virtually nothing composed prior to 1700—helped shape the classical canon. What little ink he devoted to reviewing specific performances tended to be suspicious of the flamboyant spectacle of the virtuoso. The popularity of the opera singer and virtuosic instrumentalist had grown considerably during Hanslick's career, with some performers (Paganini, Liszt) gaining cult status. This suspicion resonated with his more general distaste for the performances and compositions of such progressives as Franz Liszt and Richard Wagner (also Anton Bruckner and Hector Berlioz) and continued celebration of the performer Joseph Joachim and composer Johannes Brahms, a close personal friend.

Hanslick's critiques were rooted in his formalist musical aesthetics and the correlative irrelevance of extramusical meaning or subjects in a composition. Hanslick, of course, maintained that if music had a subject, then it was the musical form itself.[109] Here we see that he had aligned himself with philosophers and natural scientists against music theorists and aestheticians. He suggested that Rousseau, Kant, Hegel, Herbart (and Robert Zimmermann's 1865 book, *Allgemeine Aesthetik als Formwissenschaft,* founded on Herbart's principles), Kahlert, Lotze, and Helmholtz all supported his position that music had no subject. Music, as organized sounds, was self-sufficient. It did not need to express or elicit emotions through some sort of language. Recall that it was in the form itself that the musical beauty resided. This strictly autonomist position clashed with Wagner's extreme heteronomism. Hanslick was deeply troubled by Wagner's theory of opera as *Gesamtkunstwerk,* fearing that the hybridization of poetry, music, and theater implied the complete dissolution of music. The mobilization of music solely as a medium for dramatic expression was, for Hanslick, a "musical monstrosity."[110]

Friedrich Nietzsche would later exclude Wagner from the history of music, placing him in "the *emergence of the actor in music,* a capital event that invites thought, perhaps also fear."[111] In his 1888 work, *Der Fall Wagner: Ein Musikanten-Problem* (The Case of Wagner: A Musician's Problem), he sardonically explained that Wagner's compositions offered performers redemption, for neither taste nor a voice was required, not even talent was necessary since it was in conflict with the Wagnerian ideal. All that was required was "*virtue*—meaning, training, automatism, 'self-denial'. . . . Wagner's stage requires one thing only—*Teutons!*—Definition of the Teuton: obedience and long legs."[112]

Not only did the showmanship of the virtuoso trend prioritize listeners' expectations over those of the composer but it also cultivated the wrong kind of listeners. Music historian and conductor Leon Botstein argues, for example, that Brahms's regular emphasis on the need for "proper" listening was both a generational and cultural critique.[113] Brahms's complaints about the lack of rigor in musical education and training reflect an anxiety about a twofold shift: piano-based music education and the increased opportunity for individuals to listen to music without generating it themselves weakened active musical literacy.[114] Which is not to say that Brahms did not write for an audience, but rather, that he wrote for an audience trained well enough to be able to hear a piece or read a score, and then, in turn, play it back in some form. His ideal audience could experience his musical performances as if they were playing the music for themselves. The virtuoso's performance, in contrast, demanded no such skill of his or her audience.

The visual nature of the virtuoso's performance especially irked critics. The excesses of Liszt in particular, from his facial expressions to the body-encompassing physicality of his execution to his broken concert pianos, epitomized the showmanship of the virtuosic trend. The musicologist Lawrence Kramer argues that it was Liszt as virtuoso that forced the question in this era of the relationship of visuality to music, piano performance in particular.[115] It was the visuality of musical performance that was at the core of the clash between the New German School and the group surrounding Joachim, Brahms, and Hanslick. On the one hand, the virtuosic style, emphasizing visual display, far more than previous performance styles directed the audience's attention to the performer's body. On the other hand, the style embraced by the formalists prioritized the realization of the musical work alone and sought to minimize the visuality of the performance.[116] The virtuoso, in creating visual effects to dazzle the audience, forfeited himself or herself to audience expectations rather than the demands of the composer or the music itself. So music written specifically for this type of performance, one of virtuoso effect only, was a debasement of the very art of music; it was this debasement that Hanslick, Brahms, and Joachim found so problematic about the compositions of the New German School.[117]

I think we can tie this back to Helmholtz now. As Helmholtz generated and received musical sound—when playing it for himself—it was because of his musical training, his classicist tastes, and his familiarity with the mechanics of musical instrument design that he was able to reinforce his belief in the rightness, the *naturalness*, of certain tuning systems. This type of listening, one layered with specific cultural values as well as an active effort to engage the musical experience, was only possible in an individual experienced in, well, playing music for himself or herself. It was in this way that Helmholtz was able to fully observe and comprehend the genius of the composer. Helmholtz was the formalists' ideal audience member—one that studied the score, followed along, and was able to play the piece. One that exalted the music itself, the idea within.

The Psychophysics of Playing Music for Oneself

The discussion thus far has focused on what Helmholtz heard, *how* he heard, and his ability to create what he preferred to hear. I will shift now to examine how this experience of sound contributed to his natural scientific work on sound sensation. Helmholtz's threefold interaction with musical instruments—as an instrumentalist, as a member of the liberal professional taste public, and as one familiar with musical instrument manufacture—framed his understanding of sound in terms of how it was generated and which sounds he preferred to receive. His expertise with musical instruments as well as the development of new precision acoustical instruments like the siren contributed to a new understanding of sound, musical sound in particular. Helmholtz defined a musical tone as a specific type of sound wave, the product of rapid and perfectly regular periodic motion of a sonorous body. It was an external, material phenomenon—a *sound object*, as Julia Kursell has argued.[118] It had form. Further, this sound object, when combined with others in certain ratios and rhythms, was specific to time and place; as part of a musical system, the sound object had historicity.

This sound object, say a tone generated by a bow on a string, also had its *own* temporality (figure 3.2). The object changed form from the initial strike of the bow on the string, to the even vibrations achieved as the bow was moved across the string, to the dying of these vibrations after the bow had disengaged from the string (or, if the bow was stopped on the string, an abrupt stop).

A study of sound sensation for Helmholtz was a study of the nature of this external, material, historical sound.[119] In 1856, he published "Ueber

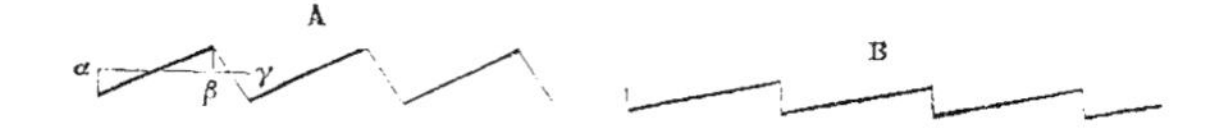

Vibrational form for the bowed middle of a violin string.

Vibrational form for the crumples that appear when the violin string is bowed between upper partial nodes.

Vibrational form for a scratching bow on a violin string—and evidence that the scratching of the bow must be understood to be irregular interruptions of the normal vibrations of the string and which makes them recommence from a new starting point.

Figure 3.2
Vibrational forms of a bowed violin string. Hermann Helmholtz, *On the Sensations of Tone*, 83–85.

Combinationstöne," the culmination of a year of research on physiological acoustics. He had spent a significant portion of this time focusing on combination tones, a well-known phenomenon where, when two musical tones are sounded simultaneously, a third tone, lower in pitch than the two sounded ones, was also audible.[120] Gustav Hällstrom had in 1832 described this third, "combination" tone as the difference between the frequencies of the original sounding tones.[121] If the sounding tones had frequencies *mf* and *nf*, where *m* and *n* are integers, then the frequency of the combination tone was $(m\text{-}n)f$. Further, Hällstrom had observed that the original sounding tones could produce more than one combination tone. That is, the original sounding tones could combine with their combination tone to generate higher-order combination tones. Hällstrom proposed that these combination tones were in fact physically identical to the beat phenomenon (buzzing or beats sounded by the wave interference of the original sounding tones). The combination tone was a beat frequency that was so rapid that it was heard as a separate tone.[122] By 1840, Hällstrom's beat theory of combination tones was supported and defended by Seebeck, Dove, Poggendorff, Roeber, and Wilhelm Weber.[123]

Helmholtz, having developed a method for generating simple tones in which harmonic overtones were almost completely absent, carried out a series of experiments substantiating an alternative explanation originally championed by Ohm.[124] With his apparatus Helmholtz could hear only one combination tone per two sounding tones, no higher-order combination tones. The higher-order combination tones must, therefore, Helmholtz argued, be a product of interfering harmonic overtones of the original sounding tones. Calling this view his "transformation theory" of combination tones, he argued that the superposition of the original sound waves could result in inharmonic distortions within the ear canal if their intensity was great enough. They were not, as Hällstrom argued, high-frequency beats heard as a separate tone. These combination tones, Helmholtz continued, objectively existed, if only in the ear canal.[125]

Historian R. Steven Turner argues that these results highlighted for Helmholtz the important role the barely audible harmonic overtones played in auditory phenomena, music in particular.[126] Helmholtz defined a musical tone as the product of rapid and perfectly regular periodic motion of a sonorous body. Just as a piano's undamped strings would sympathetically vibrate in response to a sounding tone, Helmholtz argued, so did the cortical fibers of the ear. With each fiber corresponding to a specific wavelength, the ear functioned as a Fourier analyzer, similar to the piano.[127] Now, rarely did musical instruments generate pure and simple tones as Helmholtz defined them. Nearly all simple tones were accompanied by harmonic overtones. These overtones or upper partials, Helmholtz argued, were responsible for the perceived consonance and dissonance of musical intervals. Certain intervals sounded dissonant because of beats created by the interfering overtones of the original two sounding tones. Consonance, which occurred when intervals were pure, mathematical ratios, was due to coincidence of harmonic overtones and a lack of interference-based beats. Overtones also contributed to the *timbre* of a musical tone, the difference between different instruments sounding a tone of the same pitch, for example.

The fact that this beat theory of consonance and dissonance, so dependent on harmonic overtones, predicted certain consonant intervals that happened to be the root of the Western tonal system, suggested that there was a lawlike, physiological basis for sound sensation on which musical aesthetics was based. This was at the core of Helmholtz's opus, *Tonempfindungen*. Yes, he admitted, it was surprising that such barely audible features as harmonic overtones would be so central to our experience of sound and of music. But many features of our sense organs central to sensation and perception, he continued, had long gone unnoticed by science. This was the case with harmonic overtones: "It is not enough that the auditory nerve sense the tone; the soul must also reflect upon it; hence my previous distinction between the material and *geistige* ear."[128] The coincidence of harmonic overtones, which was for Helmholtz the basis of harmonic consonance in music, was not only a physical, material phenomenon, but was also reflected on by an immaterial element of the sensory-perception apparatus. Here we have arrived at psychophysics. Sound sensation was an interpretive, empiricist process for Helmholtz. The perception of musical consonance was a product of both material and immaterial processes.

We can see here a further articulation of Helmholtz's theory of signs, which he had broadly sketched in his 1855 Kant lecture, "Über das Sehen des Menschen." Helmholtz developed the theory most fully in regard to vision, based on the idea of "local signs" proposed by Hermann Lotze and others.[129] Vision was reduced to an optimization principle. As the eye moved so that the object of observation was most clearly visible (from a specific peripheral spot to a spot of sharpest vision), the contraction of the eye muscles corresponded to a series of changing feelings of position. This series of changing feelings of position was stored in the memory and recalled and repeated whenever that specific peripheral spot was stimulated. The local sign consisted of the physical, physiological actions required to orient each spot on the retina (the direction and position of each) to the visual axis. In this way the local sign established through physical motion was associated with each point sensed and perceived in visual space; psychological sensation was due to physical stimulation.

Analogously, a listener's evaluation of their external world was not innate but instead required evaluation to relate certain timbres to certain sound sources (specific voices to individuals, instruments, etc.). Over time the mind could reason and reflect on earlier experiences from sufficient signs. In his 1868 published lecture, "Die neueren Fortschritte in der Theorie des Sehens," Helmholtz presented this as the process of "unconscious inference." This process, though it often occurred without the individual's awareness, was rational and inductive.[130] Although he did not expand this discussion into a full-fledged sign theory of hearing, Helmholtz did invoke his theory of signs to explain the perceived dominance of the fundamental tone relative to its harmonic overtones. Because a complex tone (the fundamental and its harmonic overtones) was a sufficient sign, there was no reason for the conscious mind to further analyze the tone into its constituent parts.[131]

The sign theory was also a methodological model. Just as tones existed independently of the listener to be unconsciously evaluated by the geistige ear, Helmholtz believed in a nature that could be accurately evaluated and described by natural science. Natural science could reveal the universal truths of this nature.[132] In fact, according to Helmholtz, one could only understand natural phenomena by finding the law by which natural phenomena were regulated.[133]

It is noteworthy, however, that Helmholtz connected neither his discussion of the geistige ear nor his theory of unconscious inference to the listener's ability to generate or at least reproduce sounds absent the stimulation of actual sound waves. His formalist listening would have required this ability of inner listening in order to appreciate music in the manner evident in his writings (recall that he could anticipate the difficult passages of the Mendelssohn piece), yet it remained outside the purview of Helmholtz's science. This was the case for Mach as well. As will be discussed in the next chapter, Mach depended on the ability of his readers and audiences to see notes on a page and generate the sounds for themselves in silence. It is curious that this skill was not of psychophysical interest.

To return to Helmholtz, though tone sensation was rooted in human physiology, musical aesthetics for him was historically and culturally conditioned. Put more specifically, individuals would all sense sound in the same way because tone sensation was physiological and depended solely on the anatomical structure of the ear. But the individual's endurance or taste for roughness or dissonance could vary.[134] It thus followed for Helmholtz that "the system of Scales, Modes, and Harmonic Tissues does not rest solely upon inalterable natural laws, but is also, at least partly, the result of esthetical principles, which have already changed, and will still further change, with the progressive development of humanity."[135]

Not that musical systems were entirely arbitrary. All full and perfectly developed styles of art followed well-formed rules established by artists without conscious intention, claimed Helmholtz.[136] Still it was not within the scope of the natural sciences to determine these laws. The determination of these laws was instead, because aesthetics were historically and culturally contingent, the task of historical and aesthetic inquiry. Again, Helmholtz argued, sensation and even sensuous pleasure, both rooted in the theory of consonance, were distinct from aesthetic beauty.[137]

So how could Helmholtz reconcile his theory of consonance, the basis of tone sensation, with the variable laws of aesthetic beauty? We can gain some insight from his discussion of the Arab and Persian music system.[138] Employing a modified notation to accommodate the seventeen-tone scale, he demonstrated that of the twelve tonal modes of Arab-Persian music (*Makamat*), nearly all were scales "with a perfectly correct natural intonation."[139] Comparing the Persian and Arab music to ancient Greek scales in Pythagorean intonation, he found that three corresponded, sharing natural intervals. That Helmholtz found differences between the music systems (the scales of up to seventeen tones, the use of ascending leading notes, among others, result in a decidedly different Arab-Persian musical style and sound) is not so surprising. His emphasis on the similarities is. The use of a series of fifths to establish the "perfectly correct natural intonation," Helmholtz continued, distinguished the Arab-Persian music system as noteworthy in the history of music development.[140] Music systems that adhered to natural intervals were "correct." Recall that the dominant intervals of the just intonation Helmholtz lobbied for were considered pure and natural, also "correct." From his physical, material conception of sound Helmholtz was able to develop his theory of unconscious inference to explain individuals' appreciation for pure, natural intervals: the geistige ear associated combinations of overtones with familiar voices, instruments, musical harmonies. This familiarity was the basis of a historically and culturally contingent musical aesthetics, providing the reason, for example, that an individual would appreciate the classical stylings of Beethoven or Mozart.

Unsurprisingly, through a discussion of the history of music, Helmholtz established that the modern European tonal system (read: the classical musical style) was full and perfectly developed, also "correct."[141] Helmholtz's individual, sensory interaction with musical instruments, his experience of music as a Bildungsbürger, his classicism and affiliated penchant for just intonation, were the means by which he could reconcile his lawlike physiology of tone sensation with his historically and culturally contingent aesthetics of music. The progressive evolution of the biological and the cultural could be seen intertwined in the full and perfectly developed musical aesthetics of classical Western music.

Conclusion

For Helmholtz, sound as music was not only an object of investigative study but also a *means* of investigative study. Not only were musical instruments employed as scientific instruments, music itself was an instrument of science. This music, these musical instruments, though employed in a natural scientific project, necessarily maintained their musical status and were not merely scientific instruments. And again, Helmholtz stated that he enjoyed music most when he played it himself. His individual, sensory experience with musical instruments deeply informed his work. When this individual experience of music was classical, it allowed him to reconcile his lawlike physiological theory of tone sensation with his historicist conception of musical aesthetics. Helmholtz's science was informed by the style of music he listened to and the manner in which he listened to it.

Because form was both musical structure and musical content or idea, it both transcended and was specific to history. Common to Marx's, Hanslick's, and Helmholtz's use of form is the language of figuration and materiality. To say that Helmholtz's playing a Beethoven piano sonata for himself was central to his hearing it is to state the horribly obvious. But again, his interaction with musical instruments as both performer and listener was central to his physiological theory of sound sensation. His striking of the keys sounded the piano strings, causing the cortical fibers of his ear to vibrate. Helmholtz both generated and received the sound object. He was the embodied reconciliation of his sound sensation study, his classicism, his musical aesthetics, his extensive understanding of music instrument construction, his formalist listening and performance practices —his participation in the music world generally. Pushing this further, in playing his Beethoven sonata for himself, Helmholtz was in a very real and personal and immediate way, the reconciliation of his lawlike physiology of tone sensation with his own historically and culturally contingent aesthetics of music.

4 The Aesthetics of Attention: Ernst Mach's Accommodation Experiments, His Psychophysical Musical Aesthetics, and His Friendship with Eduard Kulke

To deny the influence of pedigree on psychical dispositions would be as unreasonable as to reduce everything to it, as is done, whether from narrow-mindedness or dishonesty, by modern fanatics on the question of race. Surely everyone knows from his own experience what rich psychical acquisitions he owes to his cultural environment, to the influence of long vanished generations, and to his contemporaries. The factors of development do not suddenly become inoperative in post-embryonic life.
—Ernst Mach, *The Analysis of the Sensations and the Relation of the Physical to the Psychical*[1]

In 1863, while visiting a Viennese café, the young physicist Ernst Mach was drawn into a lively discussion among musicians over the nature of musical tones. These rowdy and informal café gatherings were a regular occurrence and Mach continued to frequent them, often presiding over affairs.[2] When later recounting his initial encounter with the group, Mach recalled that he had chosen to side with the music critic Eduard Kulke due to his more sober, more *wissenschaftlich* position on sound sensation.[3]

The others at the café were likely the circle of Viennese Wagnerians within which Kulke moved, musicians and composers, most notably Peter Cornelius, Franz Liszt, and Anton Bruckner. Kulke had at times been in contact with Richard Wagner himself, and attended the Bayreuth Festival on at least one occasion. It was, in fact, a moving performance of Wagner's *Tannhäuser* in 1854 that prompted Kulke's own "aesthetic heresy" (*ästhetische Ketzerei*). His feeling that Wagner's music was pleasing was countered by acoustic and music-theoretical arguments and accusations (think Hanslick here) that no one of taste or aesthetic education could possibly find Wagnerian music beautiful. Wagner's chords did not necessarily follow traditional rules of Western harmony and therefore could not be beautiful.

Clearly, tastes differed between individuals but the question of which taste was correct (*der richtige*), or that there could even be a correct taste, troubled Kulke greatly.[4] He could only counter by stating his own feelings, that Wagner's music was beautiful to him. The Wagnerian opera motivated a lifelong effort to reconcile the music-theoretical analyses that condemned Wagner's harmonies as ugly with his own, individual enjoyment of Wagnerian music.[5] Kulke sought to develop an aesthetic of music that legitimized both the objective analyses of acoustics and music theory and the subjective experiences of the individual.

Mach was likely interested in the discussion at Kulke's table at the Café Griensteidl because he was at the time beginning a series of experiments to locate the accommodation mechanism of hearing. Accommodation in hearing was the phenomenon in which deliberately directed attention altered an individual's aural experience. It was the listener's ability to hear, for example, the same chord differently depending on which tone of the chord the listener chose to focus on. Or—another example—it was the listener's ability to hear the cello part in a symphonic performance, again, once the listener chose to direct his or her attention to the cellos. The phenomenon had already been established and physiologically explained for vision. Mach devoted much of the 1860s and early 1870s to locating and explaining the psychophysical mechanism for the accommodation phenomenon in hearing. Though the phenomenon of accommodation in hearing was readily apparent, Mach never was able to locate the mechanism. By the 1886 publication of his opus on the psychophysics of sensation, *Die Analyse der Empfindungen und das Verhältnis des Physischen zum Psychischen* (The Analysis of Sensations and the Relation of the Physical to the Psychical), Mach had abandoned all discussion of the phenomenon of accommodation in hearing and the location of its mechanism.

Mach and Kulke's friendship and the intellectual exchange with which it began on that day in the café lasted long after Mach left Vienna, maintained through frequent visits and regular correspondence until Kulke's death in 1897. Their friendship demonstrates the extent and consequences of intellectual issues common to both psychophysics and music. Mach worked to locate and explain accommodation in hearing, the locus of an individual's subjective experience of sound. Correspondingly, Kulke sought to develop an aesthetic of music that would explain the individual's subjective appreciation of music. Theirs were parallel efforts to reconcile the individual's experience of music with musical aesthetics. An examination of the points of intersection of Mach's and Kulke's social and intellectual peregrinations reveals a very specific conception of listening that was bound to musical aesthetics. Drawing on the listening culture in which they circulated and their own personal musical aesthetics, Mach and Kulke presented an understanding of sound sensation that was psychophysical, evolutionary, and ultimately historicist, bound to time and place. So, both were developing a new understanding of listening that was psychophysical and evolutionary and therefore individual and historicist and bound to musical aesthetics.

This chapter examines some of Mach's earliest scientific work, his psychophysical studies of sound sensation. It will chart how Mach's efforts to psychophysically locate the mechanism of accommodation in hearing were deemed moot and replaced by a historical understanding of hearing. I will also document his intellectual exchange with Kulke as well as his movement in the music world more generally, both of which were central to the evolution of Mach's thinking. The two overarching goals of this chapter are to show, first, that the multivalent ways Mach, in searching for the accommodation mechanism in hearing, necessarily employed music as sound, and second, that the inseparable aesthetic aspect of Mach's psychophysical study of sound sensation fueled his historicist turn.

For Mach, it was not simply that sound and music were interchangeable, but rather, that it was critical, for both demonstrating and testing certain sound sensory perception phenomena, that he employ music as sound. So the search for the accommodation mechanism of hearing was also a study of musical aesthetics informed by his own relationship with music. Mach moderated and moved with ease through the worlds of natural science and music, actively engaging both worlds socially and intellectually. He used musical examples in his scientific writings. He wrote for musicians in musicology journals. Mach saw his work and himself as a bridge between physics and music.[6] Indeed, his project overlapped and intersected with several issues important to music theoreticians at the time.

It follows then that Mach's study of accommodation in hearing must be understood in relation to the music world. This music world was, during this period, witnessing a series of events that fueled a shift in musical aesthetics generally. This shifting aesthetics of music, especially with respect to Wagner's music, is a central part of this story.[7] There were also the significant challenges to the aesthetic traditions of the European music world of the first half of the century discussed in chapter 2, such as new tuning systems and new sounds; not even the concert A was fixed. The new tones and new harmonies—and their appeal—threatened to undermine the conception of sound sensation as universal as well as Western musical aesthetics as the most fully evolved. So as Mach and Kulke pursued their respective studies of individual musical aesthetics they did so as the music world was experiencing rather dramatic tonal destabilization.

This music aspect of Mach's psychophysical work has never been the focus of serious study by historians of science, and is in its own right quite interesting. Perhaps even more exciting, and the other main focus of this chapter, is that Mach's experiments on accommodation in hearing—and his engagement with the music world and mobilization of music to scientific ends in these experiments—contributed to the maturation of his historicist thinking. Much of the historiographical work on Mach has focused on his physics, in particular his work on the shock waves of supersonic projectile motion, and his philosophy, the fields for which he was so well respected. His phenomenology, which Mach insisted could be reconciled with experimental science, drew the ire of Max Planck and was a point of disagreement with his friend Albert Einstein. Historians of science often point to Mach's 1872 treatise, *Die Geschichte und die Wurzel des Satzes von der Erhaltung der Arbeit* (*History and Root of the Principle of the Conservation of Energy*), as his first full articulation of his position on the historicist nature of ideas —that is, that ideas were specific to time and place. This historicism directly informed the logical positivist movement that developed in the twentieth century.

Mach came to believe that the sensation of sound was not just psychophysical, not just a physiological mechanism, but also cultivated and cultured. This ultimately eliminated the need to locate a mechanism of accommodation at all, because it would constantly be changing anyway. This explains why Mach was no longer discussing the accommodation mechanism of hearing in 1885. The following examination of Mach's use of music in his study of accommodation in hearing suggests that he was thinking in a historical way, at least about sound sensory perception, much earlier than credited by historians of science. As early as 1863 Mach believed that hearing—how one heard, what one heard, what one focused his or her attention on—was bound to culture and therefore specific to time and place. How one heard could and did change over time. *Hearing itself* was historical.

Mach's Accommodation Experiments

Let us begin with Mach's accommodation experiments. From the very beginning, he relied on music. Mach usually presented the following demonstration of the phenomenon of accommodation in hearing. He urged his reader (you are invited to try this as well—see figure 4.1) to play on a guitar or keyboard instrument the chord E-G#-b-e followed by the chord A-a-c#-e.[8] Then the reader/listener was to play the chord sequence again but this time listen carefully to the higher tones (that high e). The reader/listener, according to Mach, would have the impression that the tones remained the same and only the tone quality had changed between chords. Now, playing the chord sequence one last time, the reader/listener should instead focus on the lowest tones (that transition from the low E to the low A). In this case they would hear a clear step down in pitch, as if the entire chord had dropped down significantly.[9]

This alteration of an individual's experience of sound through the deliberate alteration of his or her attention is what Mach termed *accommodation* (*die Accommodation*). Did you hear it? If you've had musical training, you likely did. It is also okay if you didn't but Mach's readers/listeners would have. Their *bildungsbürgerlich* upbringing would have guaranteed them facility on at least one musical instrument as well as music-reading ability. Which means that his readers seeing this example in print would know of the phenomenon Mach was referring to. Musical knowledge and ability were necessary for the demonstration to be meaningful. Mach's search for

Figure 4.1
Chords for performing demonstration of the phenomenon of accommodation in hearing. Ernst Mach, "Bemerkungen über die Accommodation des Ohres," *Sitzungsberichte der Kaiserlichen Akademie der Wissenschaften* 51 (1865): 343–346. This same chord example was used in Mach's 1865 lecture "Die Erklärung der Harmonie."

the mechanism of accommodation in hearing was dependent on music for demonstration, for experiment, and for argument.

From the very beginning of his study of accommodation in hearing, Mach relied on music as a means of demonstrating the phenomenon. The individual could alter his or her experience of a musical passage through an alteration of his or her attention. An individual could, for example, by focusing on certain tones, hear the opening chords of *Tannhäuser* to be quite beautiful and moving. There should be little wonder then that Mach and Kulke began such a close friendship over a discussion of musical tones that day at the café; they were working out similar problems.

A close examination of Mach's experiments on accommodation in hearing from the 1860s and 1870s reveals that Mach relied heavily on music in his science. He employed musical examples to demonstrate the phenomenon of accommodation to his audience. He also increasingly turned to music-theoretical arguments to support his assertion that the phenomenon did, in fact, exist. Until he could locate the actual accommodation mechanism of sound sensation, music was a means of demonstrating and discussing the otherwise inaccessible psychophysical phenomenon of accommodation. Music was a proxy scientific language. Mach's use of music in his psychophysical experiments on accommodation —in a time when established musical aesthetics were being called into question—in turn informed his eventual belief that hearing was historically contingent.

Mach had initially attempted to explain sound sensation in purely physiological terms. In terms of actually locating the mechanism of accommodation in hearing itself, he suspected that it was rooted in the tensor tympani and possibly also the stapedius muscles that would contract in response to altered attention, changing the transmission of sound waves to the cochlea. This model of the mechanism of accommodation began as an analogy to Hermann Helmholtz's sign theory of vision (which was very much based on Lotze's), in which the contraction of the muscles of the eye allowed an individual to spatially locate the object of his or her observation.[10] Over time, according to Helmholtz, the association of specific muscle contractions with specific locations in space became fixed and, as needed, unconsciously referred to by the individual. Similarly, Mach posited that the ear differentiated tone pitch through the contraction of various muscles in the ear (the tensor tympani and stapedius muscles) in response to actively changed attention.[11] It was this *actively* changing attention that distinguished Mach's understanding of accommodation in hearing from Helmholtz's sign theory. A psychological change (altered attention) caused a physiological change (altered sensation). Accommodation in hearing can therefore be understood as a psychophysical experience and a psychophysical investigative object.

Historian Michael Hagner sees mid-nineteenth-century psychophysical studies of attention as an indicator of the extent to which attention was redefined from earlier, late eighteenth-century conceptions.[12] Attention had previously been, he explains, a virtue, making individuals masters of themselves and the exploration of their world. The early work of Fechner, however, showed that attention was actually quite difficult to control and maintain and threw into relief the instability of the human perceptual condition.[13] Hagner points to Mach specifically and his belief that attention was a purely motor-based phenomenon.[14] He argues that this was part of a growing acceptance among psychophysicists that, through motor skills, conscious control and self-discipline influenced perception. Mach's early understanding of accommodation in hearing as purely physiological can be seen as part of this larger trend that Hagner describes. If Fechner had determined that perception changed in spite of, possibly even because of, focused attention, Mach's interest in accommodation was to determine why this happened.

As previously stated, Mach began his investigations from a physiological perspective. In his 1863 article, "Zur Theorie des Gehörorgans," he described his efforts to apply a kymographic theory of the ear to accommodation in hearing.[15] Like the kymograph, which recorded blood pressure through a stylus on a graph on a rotating band of paper, Mach believed that the ear drew (*zeichnen*) the sound waves in the fluid of the inner ear labyrinth. These sound waves were then absorbed by the auditory nerve. The entire ear—the eardrum, the three middle ear bones (ossicles), and the fluid of the labyrinth—functioned to transcribe the sound waves from the medium of air molecules to the medium of the fluid. Though the sound wave was modified (regulated and damped), no analysis occurred. Mach demonstrated all of this through a series of experiments and mathematical derivations.[16]

One implication of this kymographic theory that interested Mach was the simultaneous reflection of the sound waves transmitted by the eardrum, an analog to Kirchhoff's theorem of the equal absorption and emission of light waves. Mach performed a series of experiments demonstrating this effect.[17] Placing an assistant with a long rubber tube in his ear in another room, Mach very softly sang a constant tone while moving the other end of the tube back and forth, relative to his own ear. The tone was loudest, according to the assistant, when Mach's end of the rubber tube was nearest his (Mach's) ear, when the sung tone was amplified by the reflection of its sound waves in Mach's ear. In another experiment, Mach softly sang a tone with an end of the rubber tube in each of his own ears. When he pinched off the tube in the middle with his fingers he noticed a decrease in the volume of the sung tone, presumably because his pinching had eliminated the sound waves being reflected back and forth between his ears through the tube.[18]

Thus Mach believed he had both theoretically derived and experimentally demonstrated the mechanics of his kymographic model of the ear, the transmission of sound waves through the ossicles to be transcribed into the fluid of the labyrinth. But the model could not necessarily explain the ability of the listener to actively distinguish a single tone as distinct from other tones sounding simultaneously. It could not explain accommodation in hearing. The musical examples clearly demonstrated the phenomenon of accommodation and yet Mach could not locate and directly observe the mechanism of accommodation. Still, he maintained his belief that attention was a bodily function and that, therefore, the phenomenon of accommodation had its foundation in the mechanisms of the body.[19] So he continued with his search.

Elaborating on these early investigations, Mach performed a series of experiments in the summer of 1863 with Joseph Popper and students of the Vienna Physical Institute. For these experiments Mach placed a vibrating tuning fork in his teeth and one end of a rubber tube in one of his ears.[20] The other end of the rubber tube was placed in an assistant's ear. As the tuning fork sounded Mach slowly changed his attention from the fundamental, or ground, tone to various harmonic overtones. Mach could hear these overtones as strong and distinct from the ground tone as he moved his attention from one overtone to the next. But the assistant could not. Although other experimental work electrically stimulating the tensor tympani had established the muscle's ability to change the tension on a prepared (nonliving) eardrum, and although Mach had mathematically demonstrated that changed tension on the eardrum would result in higher tones appearing louder, his tuning-fork experiment could not confirm that changed attention correlated with changed eardrum tension in turn correlated with changed sound sensation (hearing the overtones more strongly).[21] He was left to conclude that while his kymographic theory held promise, further experimental proof was required to show that it explained the phenomenon of accommodation.[22]

If the use of music to demonstrate the phenomenon of accommodation was an early indicator, in 1865 Mach began to further integrate music into his work on sound sensation. He presented two popular lectures on musical acoustics, "Über die Cortischen Fasern des Ohres" and "Die Erklärung der Harmonie." In both, his discussion of the accommodation mechanism was in musical terms. The first, "Über die Cortischen Fasern des Ohres," introduced the concept of sympathetic vibration. He described this as the sounding of a sonorous body when, either alone or in the company of others, its special note (*sein Eigenton*) was struck.[23] Groups of sonorous bodies behaved similarly—individual bodies within the group only sounded when their particular note was struck. Replicating a demonstration introduced by Helmholtz, Mach turned to the piano as an example of such a group of sonorous bodies.[24] Standing two pianos next to each other, lifting the dampers on one piano (by pressing the sostenuto pedal), and striking a key on the other resulted in the same note ringing on the undamped piano. Similar results occurred in response to a major triad, and so on. The undamped piano, Mach explained, separated the sounded tone in the air into its individual component parts, a spectral analysis of sound.[25]

Mach used the piano as an analogy for both the mechanism and the phenomenon of accommodation in hearing. According to Mach, the ear had the same ability to perform a spectral analysis of sound. Mobilizing the same analogy employed by Helmholtz, Mach directed his audience to imagine the undamped piano as, instead, the cortical fibers of the ear. Just as a string on the undamped piano would sound in response to a tone sounded by the other piano, so too would a single cortical fiber be thrown into vibration. The large number of cortical fibers—one per piano string—allowed for accommodation and thus an appreciation of music. A listener could, for example, pick out the melodic lines of a Bach fugue. Or they could distinguish separate tones of simultaneous impressions (*Eindrücke*), not just a harmonious chord but any combination of tones, by directing their attention.[26]

In "Die Erklärung der Harmonie," Mach elaborated on the role of attention in an individual's experience of music. Again he turned to the piano to illuminate his point. An individual would hear the harmonic sequence of two different chords voiced the same (all tones the same loudness) in succession differently depending on which tones—the roots of the chord that changed or the upper tones that remained unchanged—they directed their attention to (this was the exercise demonstrating the phenomenon of accommodation discussed earlier). It was therefore the art of music composition, Mach asserted, to guide the listener's attention, through, for example, voicing portions of the chords to bring out certain tones over others. There was also an art of hearing, he continued, which was not the gift of every person.[27] Only through extensive practice could one develop the ability to further differentiate a single tone into its fundamental tones and harmonic overtones. These overtones, Mach claimed, played an important part in the formation of musical timbre as well as the consonance of sound, a clear reference to Helmholtz's work on tone sensation.[28] Attention combined with the accommodation mechanism allowed the individual to distinguish harmonic overtones, the root of Western harmony. Mach's work suggested that Western musical aesthetics both shaped and were the product of the accommodation mechanism of hearing.

Mach used musical examples to support his arguments in technical articles as well. In the 1865 publication in *Sitzungsberichte der Kaiserlichen Akademie der Wissenschaften*, "Bemerkungen über die Accommodation des Ohres," Mach presented musical examples of chord progressions (the same as in "Die Erklärung der Harmonie" and "Die Cortischen Fasern des Ohres") and of melodic lines to demonstrate the phenomenon of accommodation. Although he sought to locate the accommodation mechanism psychophysically, he used music to best illustrate the phenomenon for his audience at the Akademie der Wissenschaften.

The pervasive use of musical examples and arguments that Mach employed in these discussions of the accommodation mechanism suggests that he made the following three assumptions. First, he assumed music was a valid avenue through which to understand the sensation of sound because music and sound were equivalent or at least closely related. Second, he thought his audiences were well versed in music theory and had extensive musical experience —enough to follow and be convinced by the musical passages written on the page. Third, Mach's frequent use of musical examples also assumed the validity of his audience's individual experience of music. An experience of the accommodation phenomenon could only be subjective. Mach was pursuing a resolution to Kulke's dilemma of developing a musical aesthetics that reconciled acoustic and music-theoretical analyses with individual tastes. By attempting to root sound sensation in accommodation, Mach tied it to the attention of the individual and his or her experience. Individual attention was of course part of individual aesthetics, and specific to local time and space. It was historicist, at least for a single individual.

In the early 1870s Mach abandoned his attempts to locate the accommodation mechanism through the direct observation of the accommodation phenomenon in himself and his assistants. Instead he attempted to recreate the phenomenon on a nonliving, prepared ear. The experimental preparation was as follows (see figures 4.2 and 4.3). First he carefully removed the tensor tympani in order to attach a thread to it. Once it was reinserted into the middle ear cavity, weight on the order of a few grams could be

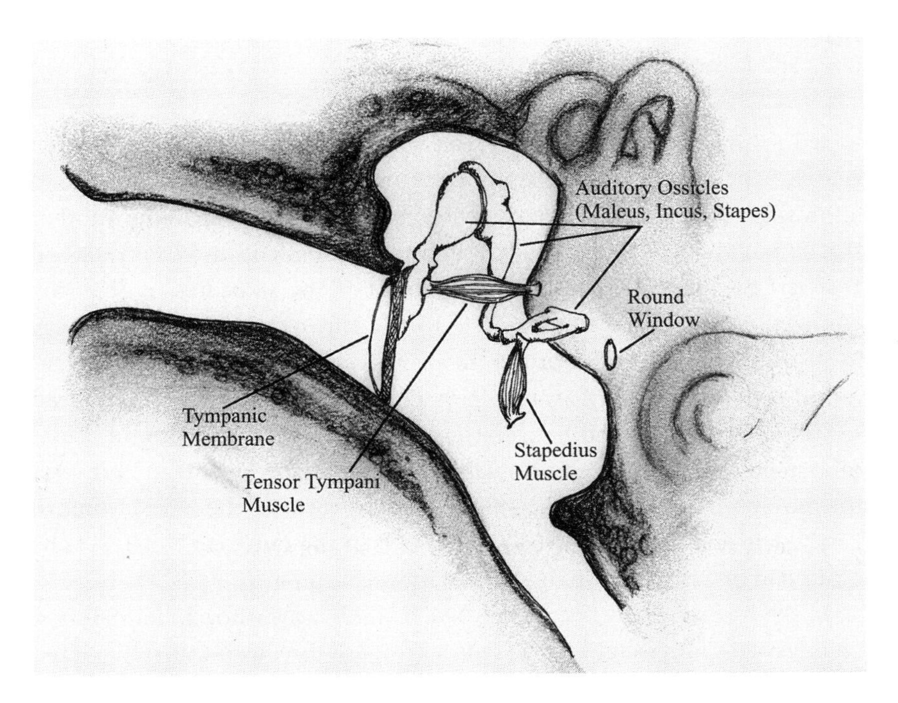

Figure 4.2
The middle ear

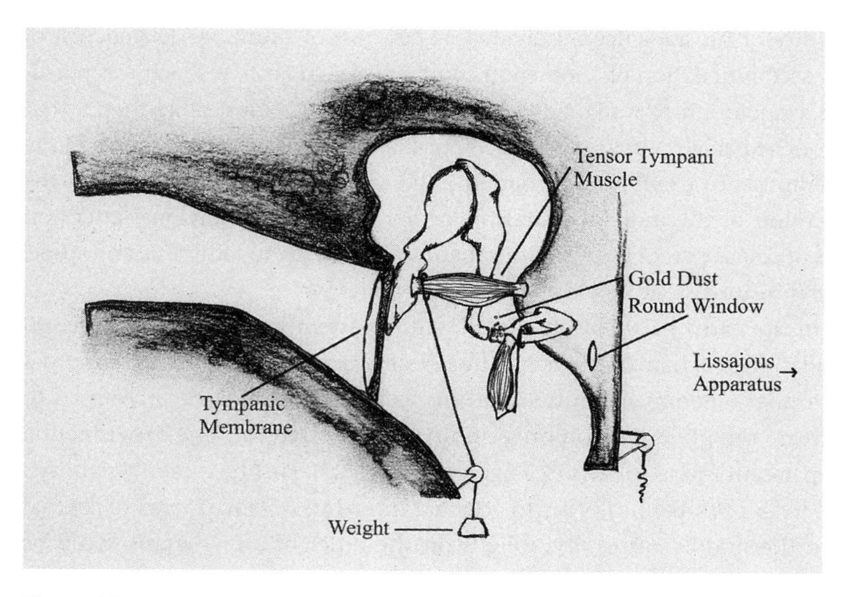

Figure 4.3
Mach and Kessel's prepared ear

added to the other end of the thread in order to produce tension in the muscle. The lower, bony portion of the middle ear cavity was cut out to allow space for the thread to pass over a pulley and hang vertically. Another thread was attached to the head of the stirrup bone. This thread was passed through the stapedius muscle canal alongside the stapedius muscle (the connection of the stapedius muscle to the stirrup bone was left intact). Most of the bony portion of the inner ear—everything past the oval and round windows—was cut away to provide space both for the stirrup thread to pass over a pulley and hang vertically and for a microscope and ocular micrometer to measure the displacement of the stirrup thread and view the movement of the gold-flake-dusted ossicles through the round window. Lastly, one end of a rubber tube was placed in the outer ear canal with its other end attached to the opening of an organ pipe.

In the first series of experiments Mach and his coauthor, Johann Kessel, attached a 3-gram weight to the tensor tympani thread and then measured the displacement of the stirrup thread, which correlated with the movement of the ossicles, when an organ pipe of 256 cycles per second was sounded. This created tension on the eardrum and constrained the movement of the ossicles, approximating the assumed influence that attention would have on the tensor tympani. When they ran the experiment again with an organ pipe of 1024 cycles per second they found that the stirrup thread displacement was less. When they pulled on the stapedius muscle, further constraining the movement of the ossicles, the displacement of the stirrup thread was reduced further. Mach and Kessel believed that this demonstrated that changed tension on the eardrum resulted in changed transmission of the sound waves through the ossicles.[29]

Next, they attached both organ pipes via rubber tubes to the single rubber tube in the outer ear canal. A Lissajous vibration microscope was also attached to the system and set up to project two-dimensional images of the stirrup thread displacement.[30] First, with no weight on the tensor tympani thread, the low pipe was sounded. The image projected by the vibration microscope can be seen in the first of the set of figures (see figure 4.4). When the same pipe was sounded while there was tension on the tensor tympani (weight was added), the second image was seen. When just the higher pipe was sounded, with no weight on the tensor tympani, the third image was seen. When both pipes were sounded simultaneously, again with no weight on the tensor tympani, the fourth image resulted. They then tried sounding both pipes and weighting the tensor tympani. The image that resulted was the third, the same image as when the higher pipe sounded with no weight. Changing the tension on the tensor tympani while two tones were sounding simultaneously altered the movement of the ossicles to appear as if only the higher tone was sounding. It appeared that Mach and Kessel had located the accommodation mechanism, that it was as suspected, the tensor tympani.[31]

That same year Mach and Kessel published another article discussing more experiments on the middle ear, a summary of their efforts to determine

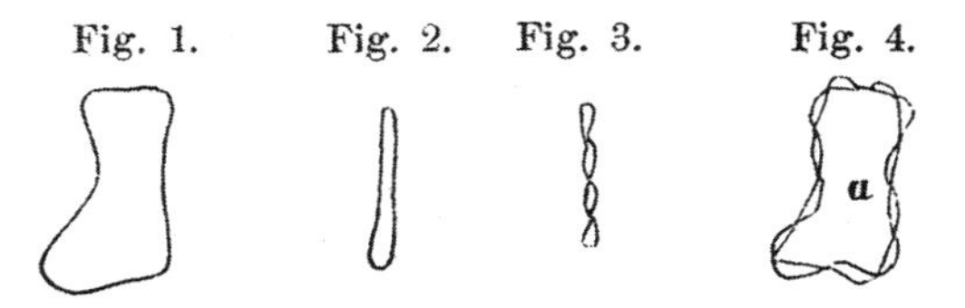

Figure 4.4
The Lissajous vibration microscope images. Ernst Mach and Johann Kessel, "Versuche über die Accommodation des Ohres," 339.

the role of the middle ear cavity (*der Trommelhöhle*) and the Eustachian tube, ultimately the role of air pressure difference, in sound sensation.[32] One experiment in particular further reinforced their earlier work on the prepared ear. It consisted of placing an observer in a box airtight enough for the pressure to be noticeably increased or decreased. The observed tension on the eardrum due to the change in pressure and then eventual equilibration indicated that the Eustachian tube was usually partially closed.[33] A lower tone sounded during increased tension on the eardrum appeared to the observer inside the box to dissipate faster than a higher tone. Kessel even claimed, for a single sounding tone, to hear the fading of the fundamental tone and the amplification of the overtones. He also claimed to hear better when the box was at negative pressure relative to the outside air. Though Mach and Kessel seemed tantalizingly close to locating the accommodation mechanism, their psychophysical investigation remained inconclusive.

They then attempted to replicate this series of experiments on a living ear. They of course could not cut away bone in order to attach little pulleys and weights on a living person and so instead constructed an "ear mirror" (*ein Ohrenspiegel*) with which to observe the displacement of gold flakes on the outside of the eardrum. The vibration microscope images for the displacement of the eardrum of the living ear for lower tones were the same as those for the prepared ear. They were not, however, the same for higher tones. Mach and Kessel could not recreate the mechanism of accommodation in a living ear. This discrepancy between results on a nonliving ear and a living one, Mach claimed, indicated that listening and hearing were not the same thing. Attention and accommodation could not be replicated by merely adding a weight and pulley system to the tensor tympani.[34] The psychophysical nature of accommodation in hearing limited its study to individuals' subjective observation of the phenomenon. The mechanism of accommodation in hearing could not be located by outside observers. But if Mach could not locate the accommodation mechanism through psychophysical experimentation, perhaps he could through music.

Mach's Musicological Writings

Mach's use of music in his search for the accommodation mechanism in hearing was not simply a consequence of the musical milieu. He deliberately employed music in his psychophysical experiments not simply as sound but as music. It was critical that he use music as music to study accommodation in hearing. First, music was the best way to convincingly demonstrate the existence of the phenomenon of accommodation. And, in turn, locating the accommodation mechanism had the potential to buttress efforts within musicology—such as Kulke's—to explain individual variation in musical tastes. In his search for the mechanism of accommodation in hearing, Mach did not simply borrow music for his psychophysical experiments. He simultaneously offered up his findings to the music world. In the two decades during which Mach studied the phenomenon of accommodation in hearing he was also writing about sound sensation more generally for a musical audience, in musicology journals and encyclopedias. Further, Mach was in regular communication with Kulke about his work because he too was struggling with issues surrounding the individual's subjective experience of sound.

Mach actively reached out to the music world with the results of his research on sound sensation. He was not just integrating music into his work, but also himself into music. In an early-1860s letter to Kulke about the music critic's interest in evolutionary theory, Mach noted their role reversal: "So we will both perform it this way. You as musical physicist. I as a musician."[35] Throughout the 1860s and 1870s Mach wrote articles for music and musicology journals, and for the twelve-volume music encyclopedia *Musikalisches Conversations-Lexikon: Eine Encyklopädie der gesammten musikalischen Wissenschaften*.

By 1866 Mach was self-consciously placing himself at the intersection of the natural scientific and music worlds with his book *Einleitung in die Helmholtz'sche Musiktheorie: Populär für Musiker dargestellt*. It was a reworking of Helmholtz's much heralded 1863 treatise, *Die Lehre von den Tonempfindungen*, in terms a musician could understand. This meant that physics would be discussed only when absolutely necessary, and there would be no descriptions of experiments, no mathematics. With only a few exceptions, Mach simply summarized Helmholtz's main points. One significant exception, however, was Helmholtz's understanding of attention's role in sound sensation as, according to Mach, a strictly psychological sense. Mach, of course, believed attention worked in conjunction with accommodation to cause a physiological change in the ears in relation to individual attention.[36] But overall Mach was supportive of Helmholtz's work.

He hoped his contribution would lead to cooperation between musicians and physicists and that it would help some musicians overcome their initial reticence about studying Helmholtz's work.[37] Certainly individuals who were both physicists and musicians would find the publication easy to follow.[38] But even for others, once the bridge of ideas between physicists and musicians was constructed, Mach believed it would continue to develop.[39]

An 1867 review of Mach's book in the *Leipziger Allgemeine musikalische Zeitung* claimed to welcome any work that would help the musician befriend Helmholtz's research through further illumination.[40] The author hesitantly allowed that it remained to be seen, however, whether Mach's survey would provide such illumination, especially the section on harmony. Three months later the same journal ran an extended article titled "Zur Theorie der Musik: Die Physiker und die Musiker" that explored the overlapping intellectual circles of Helmholtz and Hauptmann's physiological acoustics with music philosophy. A third of the article was devoted to a discussion of Mach's reworking of Helmholtz. The editors recommended Mach's publication to anyone hoping to gain insight into Helmholtz's theories.[41] But they also explained that, while Mach's work was both instructive and convincing, it missed Helmholtz's main points. These main points were, apparently, that Helmholtz's "physical-psychological" theory was internally consistent and that this theory was in full harmony with the art of music.[42] So in 1867, although the musicologists cautiously welcomed Mach's presence in the music world, they were also willing to question and challenge his theories.

Eleven volumes of the *Musikalisches Conversations-Lexikon*, edited by Hermann Mendel, were published between 1870 and 1879, with a supplemental volume in 1888. The other contributors were described as music critics, composers, music theorists, and music directors, many located at the Leipzig Conservatory. The articles were short, rather straightforward entries intended to be both *wissenschaftlich* and approachable for a lay audience. They were also unsigned so it is not possible to determine exactly which ones Mach authored. The articles for "Ohr" (ear) and "Tonempfindungen" (tone sensation) both referred extensively to Helmholtz but the latter in particular contained hints of Mach's thinking. The ear was described as a mediator between complex sound waves and tone sensation, the first assumption of the tone art (*Tonkunst*). Further, the pleasure of music was a result of not only the sensation of tones but also the embodiment of meaningful works of art.[43]

Mach's participation in the *Musikalisches Conversations-Lexikon* indicates the extent to which his prestige had grown since his initial efforts to engage the scholarly music community with his work on Helmholtz. Mach was now regarded as an expert. The "Mach" entry in the *Musikalisches Conversations-Lexikon* described him as one of the "most worthy scholars of the science of music." His book *Einleitung in die Helmholtz'sche Musiktheorie* was declared extraordinarily worthwhile.[44] In 1887, one of the leading musicology journals, *Vierteljahrsschrift für Musikwissenschaft*, published a lengthy excerpt from Mach's *Die Analyse der Empfindungen*. Mach's efforts to construct a bridge of ideas between physicists and musicians appeared successful. Certainly a direct exchange of ideas is seen in Mach and Kulke's respective projects. The next section examines the extent to which they drew on similar evolutionary perspectives with regard to musical aesthetics. While Kulke's effort was, arguably, a simplistic dead end that offered little absolution to his aesthetic heresy, Mach's evolutionary approach was the remaining element necessary for the development of his belief in the historicist nature of hearing.

Mach and Kulke's Psychophysical Evolutionary Theories

As previously mentioned, Kulke, like Mach, aimed to reconcile the individual experience of sound/music with musical aesthetics. Mach and Kulke also shared an understanding that this individual experience of sound was not only psychophysical, but also evolutionary. Ultimately, both came to believe that hearing itself, like musical aesthetics, was historicist, bound to time and place.

It must be noted that there was a distinctive racialist discourse threading its way through music circles at this time, especially prominent in the large cities of Central Europe.[45] Much of this was fueled by the introduction of a scientific concept of race hierarchy, possible through the application of Darwinian evolutionary theories to culture, and successful because it reinforced preexisting cultural assumptions about race and class. The music traditions of Jews and Gypsies had long been dismissed as the menacing product of Orientalized "others."[46] With increasing nationalism and anti-Semitism in the second half of the nineteenth century, this sense of menace became sharper and more shrill. This growing anti-Semitism, in turn, alarmed the growing Zionist movement led by Theodor Herzl. The writings of Wagner, most notoriously his deeply anti-Semitic 1850 pamphlet *Das Judenthum in der Musik*, written under a pseudonym and republished under his name in an expanded form in 1869 as *Aufklärung über das Judenthum in der Musik*, suggested that the proponents of autonomism in music were nothing more than a conspiratorial cabal devoted to promoting "Judaized" music culture (especially the compositions of Mendelssohn).[47] The virulence of the racialized, anti-Semitic rhetoric in relation to music, invoking evolutionary theory as needed, cannot be ignored as part of the cultural context in which Mach and Kulke operated. Kulke identified as Jewish, and so an awareness of Wagner's writings, in particular, adds an additional layer to Kulke's claims of aesthetic heresy because of his immediate attraction to *Tannhäuser* in 1854.

Although Mach had read Darwin's *Origin of Species* soon after its publication in 1859 and later integrated many elements of it into his theory of "economy of thought," his conception of evolution was very much rooted in Lamarckian inheritance of acquired traits, as was typical at the time. Mach had read Lamarck first. The evolutionary theory Lamarck initially proposed in 1809 was one of parallel linear progress of forms (no common ancestor) from spontaneous generation. Echoing elements of the temporalized Great Chain of Being, Lamarck conceived of this linear progress as a progressive trend from lower forms of life to higher levels of organization. He believed that evolutionary change was primarily driven by "wants," a combination of both desires and needs. Use or disuse would lead to the development or loss of a physiological trait.[48] This physiological change could then be passed on from parent to offspring, and over many generations the modification would eventually be reified as a permanent physiological change. Lamarck considered this principle of the inheritance of acquired traits to be a secondary factor in his evolutionary theory.[49]

And yet it was this secondary law of inherited acquired traits that gained a new popularity—in isolation from his full program—during the latter half of the nineteenth century.[50] Neo-Lamarckism was a diverse movement, varied both geographically and disciplinarily, but unified by the perceived shortcomings of the Darwinian evolutionary mechanism of natural selection. Neo-Lamarckians appreciated that the inheritance of acquired traits was directed by the individual organism's wants rather than the whims of natural selection. In this sense neo-Lamarckism was appealing because it was moral; organisms were not at the mercy of their environment and could retain control over their own destiny.[51]

An indication both of the popularity of neo-Lamarckism and of the common conflation of neo-Lamarckian and Darwinian evolution can be seen in the work of Herbert Spencer, among others. Spencer invoked a combination of natural selection and Lamarckian inheritance of acquired traits to explain the mechanism of evolution. The inheritance of acquired traits, which he conceived of as a process of individual self-improvement, was central to his work and he vigorously defended Lamarckism against attackers. Ernst Haeckel as well, though he called himself a Darwinian, only applied natural selection in his evolutionary theory at the species level, as a means of eliminating the less successful products of direct, Lamarckian adaptation. So Darwinian and Lamarckian conceptions of evolution were often tangled at the end of the nineteenth century. Individuals that described themselves as Darwinians—or have since been described as Darwinians, like Herbert Spencer—frequently gave Lamarckian inheritance equal if not greater weight in their evolutionary perspectives.

Mach credited Ewald Hering's work on the inheritance of memory as the most valuable contribution, motivating him to expand his understanding of evolutionary inheritance to include psychological traits. Hering originally presented "On Memory as a General Function of Organized Matter" before the Imperial Academy of Sciences in Vienna on May 30, 1870.[52] Hering elaborated on Lamarckian inheritance to include Gustav Fechner's psychophysics.[53] He argued that the inheritance of acquired traits included not only physical but psychical traits—they were, according to Fechner, the same after all. Memory, Hering claimed, was the unifying force of consciousness. It was, in powering the force of reproduction, the primary force of organized matter. All organized beings were a product of the unconscious memory of organized matter, which, ever increasing and ever dividing itself, ever assimilating new matter and returning it in changed shape to the inorganic world, ever receiving some new thing into its memory, and transmitting its acquisitions by the way of reproduction, grew continually richer and richer the longer it lived.[54]

An inherited memory germ that was Lamarckian in essence generated the development of individual beings. Memory as the universal function of organized matter, as Hering asserted, implied that "the development of one of the more highly organized animals represents a continuous series of organized recollections concerning the past development of the great chain of living forms."[55] Yet organisms were not mere automata motored by blind instinct. Instinct was, Hering claimed, inherited memory of nerve substance actively accessed by the organism. Thus, every instinctive reflex was in fact a deliberate act of recollection.[56] This allowed Hering to maintain both his nativist position, asserting that perception was innate, and one in which the organism was an agent over its own destiny.

Hering's ideas about the transmission of psychical traits can be seen echoing in Mach's discussions of the development of knowledge. By the 1880s Mach was describing the humble beginnings of knowledge as the half-conscious (*halbbewusst*) and automatic (*unwillkürlich*), "instinctive habit of mimicking and forecasting facts in thought, of supplementing sluggish experience with the swift wings of thought at first only for their material welfare."[57] Mach claimed that these "primitive psychical functions" (*ersten psychischen Functionen*) were rooted in the "economy of the organism" (*der Oekonomie des Organismus*), like the sense of motion or digestive abilities, and that such acts formed the basis of current scientific thought—our instinctive knowledge.[58] Humans as a species gained their knowledge through the experience both of individuals and of the species as a whole. Successful communication of this knowledge allowed for experience to be preserved and inherited by subsequent generations. It was this successful communication, this "economy of thought," that became Mach's central goal for science—"economically arranged experience."[59]

Mach had understood aesthetics to be evolutionary from very early on. In 1867 he had presented a lecture titled "Why Has Man Two Eyes?," in which he first addressed the question as a physiological one, explaining that the two eyes were required for depth perception. But his discussion soon turned to a survey of the different visual aesthetic traditions of ancient cultures. Mach stated: "Change man's eye and you change his conception of the world. We have observed the truth of this fact among our nearest kin, the Egyptians, the Chinese, the lake-dweller."[60] He was suggesting that changing physiology—the changing of the sight organ itself—explained the variety of visual aesthetic traditions throughout time and place.

In 1871, while he was particularly engrossed in his search for the accommodation mechanism in hearing, Mach wrote "Ueber die physikalische

Bedeutung der Gesetze der Symmetrie," outlining his belief that repetition of sensory stimulation was the key to "agreeableness."[61] Symmetry was the most agreeable because it conditioned repeated sensations, most noticeable in the visually pleasing effect of regular figures, especially straight vertical and horizontal lines. Mach devoted quite a bit of the article to an attempt to locate symmetry in sound sensation. As discussed in the introduction to this book, he performed a series of experiments in which he played a piano with the sheet music reflected vertically and then horizontally in a mirror. The melody became unrecognizable. Harmonic intervals, when mirrored, were reversed (a series of major key intervals sounded in a minor key on the mirrored piano and vice versa).[62] Because sound sensation was temporal, rather than spatial, as in vision, Mach concluded that there was no symmetry in the province of sounds other than an intellectual one in the transposition from a major to a minor key in the Western harmonic system.[63]

Still, Mach continued, human appreciation for symmetry was not simply due to the physiology of the sensory organs but instead deeply rooted in the organism itself.[64] He claimed that human notions of beauty could very well be different if their physiology were different. With that, Mach asserted that conceptions of beauty could be modified by culture, "which stamps its unmistakable traces on the human body."[65] The idea of eternal beauty was thus mistaken; had not all musical beauty been restricted to a five-toned scale in the recent past?

So, aesthetics were psychophysical. And an individual's aesthetics were subject to Hering/Lamarckian inheritance. In fact, because culture "stamps its unmistakable traces on the human body," musical aesthetics was an alternative entry point to investigating sound sensation. Already in 1867 Mach was thinking of the sense organs as well as visual aesthetics within an evolutionary, historicist framework. Certainly by 1871 he was. Further, there are indications that he was thinking in this manner as early as 1863—the same year he met Kulke and began his search for the mechanism of accommodation in hearing.

So it is likely that Mach understood this inherited knowledge to be quite broadly defined, that it extended to both the psychophysics of sensation and aesthetics. In an 1872 letter to Kulke, Mach asked if they could presently hear what the Greeks had heard and the Slavs still hear. And could the answer lie in attention (*Aufmerksamkeit*) only? He mused that a developmental history of melody, harmony, and rhythm would be very interesting.[66]

In a later letter Mach asked why the Germanic and Slavic peoples phrased their melodies differently. Kulke replied that the question was a historical one.[67] It was also, Kulke continued, an issue of aesthetics and a further application of Darwinian evolution to the arts. For both Mach and Kulke, musical aesthetics were subject to variation and inheritance much like physiological attributes.[68]

Kulke later more fully elaborated his "Darwinian" (Lamarckian) theory of melody in *Über die Umbildung der Melodie: Ein Beitrag zur Entwickelungslehre*.[69] He had apparently written much of the manuscript in the 1860s but it was only after extensive prodding by Mach that it was published in 1884. While Kulke described his work as an application of "Darwinian" theory, he was also, perhaps only, thinking of the Hering/Lamarckian theory of inheritance. Clearly accepting of the theory of the inheritance of ideas, even of collective cultural memory, Kulke wanted to explore the possibility of applying evolutionary theory to musical melody.

He presented a demonstration of how, when reduced to its dotted-sixteenth note motif, the Andante movement of Beethoven's Symphony in C Minor (No. 5) was also the motif of a traditional country dance (*einen Bauerntanz*).[70] (See figure 4.5.)

Kulke then further reduced this passage to the theme in figure 4.6.

Thus reduced, Kulke noted that the theme was in fact a compound of two different components—physiological and psychological—but he did not expand on this point.[71] But it should be noted that he understood

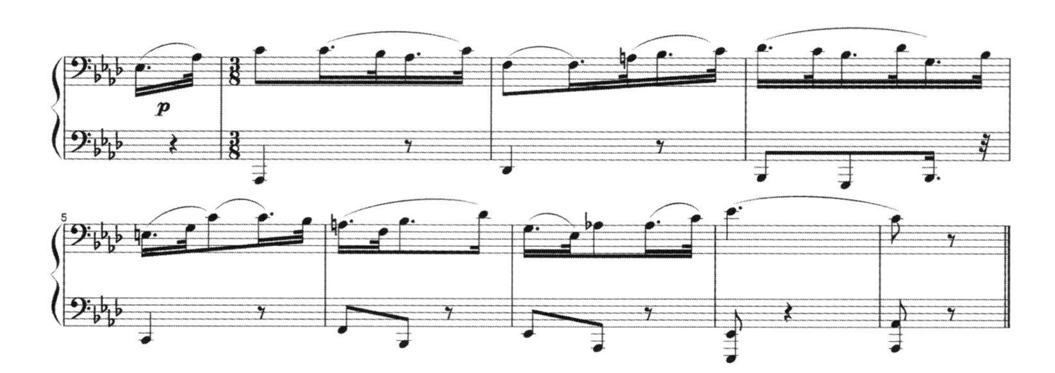

Figure 4.5
Theme from Andante movement of Beethoven's Symphony in C Minor. Kulke, *Über die Umbildung der Melodie*, 4.

Figure 4.6
Further abstracted theme from Andante movement of Beethoven's Symphony in C Minor. Kulke, *Über die Umbildung der Melodie,* 5.

Figure 4.7
The theme of the Andante movement of Beethoven's Symphony in C Minor reduced to its highest simplicity, the dotted-sixteenth. Kulke, *Über die Umbildung der Melodie,* 6.

Figure 4.8
Kulke's example of the Bauerntanz motif similar to theme of Andante movement of Beethoven's Symphony in C Minor. Kulke, *Über die Umbildung der Melodie,* 7.

musical aesthetics to be psychophysical. He then proceeded to address the repetition of the two-bar theme of the first two (complete) bars, which was paralleled in the third and fourth bars. A further reduction of the theme eliminated the strict symmetry but maintained the dotted-sixteenth motif.[72] This dotted-sixteenth could then be extrapolated out for the full eight bars, shown in figure 4.7.

In this form, Kulke argued, the Andante theme was also the Bauerntanz motif. (See figure 4.8.)

Kulke performed a similar demonstration with Mozart's Piano-Violin Sonata in G Major and Lanner's "Pester Walzer." His work had, he believed, validated the application of Darwinian evolution not only to *geistige* structures (not just physiological and behavioral) but to melody in particular.[73] One can also see elements of a Hering/Lamarckian conception of the inheritance of acquired ideas in Kulke's work. The melodies—the ideas—themselves were the individual organisms undergoing evolution. Certainly Kulke's work reflects a historicist conception of musical aesthetics, but by so explicitly articulating musical aesthetics in evolutionary terms (both Darwinian and Hering/Lamarckian) it is clear that he also understood musical aesthetics to be psychophysical.

Kulke's analyses were, admittedly, a bit clumsy, because he seemed to believe that by showing similarities of rhythm and melody he was showing evolution. By all appearances, Kulke simply reduced famous melodies to rhythmic approximations of older, traditional ones. This certainly suggested that some sort of melodic and harmonic development had occurred, but did not provide an explanation of how or why this development took place. Kulke had in fact presented very little that appears Darwinian (no discussion of a mechanism of natural selection, for example), but his efforts are notable for his goals if not his success in achieving them. And it is significant that Kulke repeatedly referred to his efforts as the application of evolution to melody and that, ultimately, he understood musical aesthetics to be evolutionary.

It is necessary to briefly mention the Viennese music theorist Heinrich Schenker. Born in what is now Galicia, he went to Vienna originally to study law but also enrolled at the conservatory to study piano with Ernst Ludwig and harmony with Anton Bruckner. In addition to having a modest career as an accompanist, he also edited and wrote music criticism. Evident in both Schenker's editing and analytical writings was his effort to understand and respect the compositional process. He insisted that both performers and listeners abide by composers' intentions. In his 1894 essay "Das Hören in der Musik," Schenker—much like Hanslick and Riemann—criticized the superficial listening that had grown prevalent among contemporary listeners.[74] He recommended repeated and disciplined listening in order to grasp the composers' intentions. While much of his music theory consisted of lengthy discussions of musical perception, Schenker was unconvinced that science could provide much insight into musical aesthetics, at times making both oblique and direct references to Helmholtz's *Tonempfindungen* in his writings.[75] Or, to take another example combining his skepticism toward science with xenophobia: Schenker nostalgically embraced the Vienna-based Bösendorfer piano (which Mach also played) as an alternative to the Steinway, which he saw as a foreign invader embodying precision engineering and scientific principles.[76]

Schenker was acquainted with Hanslick, friends even. Though there were certainly clear differences, musicologist Nicholas Cook convincingly finds many affinities between Hanslick's and Schenker's conceptual frameworks. Both sought to understand emotional expression in music in terms of its objective properties.[77] Cook claims that much of Schenker's work was an effort to refine Hanslick. Schenker similarly derided the music of the New German School, instead celebrating Bach, Schubert, Mendelssohn, and Brahms for their ability to transmute their subjective feelings into musical objectivity.[78]

In this understanding of compositional skill, Schenker's language echoed that of Riemann. Though Schenker saw Riemann as his chief rival, they shared a contempt for contemporary music. Both understood that there was a moment of transmutation of the subjective into the objective in a musical performance. For Schenker it occurred during the composition process, as the composer created a musical object. For Riemann it occurred during the listening process, as the careful listener either subjectivated music to his or her will or objectivated it to the imagination of the composer.

I bring up Schenker because it was a decade after Kulke's publication that Schenker presented his explanation for the origins and cultural variations in music. In his early 1895 article, "Der Geist der musikalischen Technik," he presented an origin story or a story of the taming of the aimless and self-indulgent musical impulse by language and laws.[79] Schenker was communicating with Mach about his work at this time, possibly about "Der Geist der musikalischen Technik," and Mach generally agreed with his opinions and was nice enough to give him the contact information for the Viennese comparative musicologist Wallascheck.[80] Later Schenker's rhetoric became much more naturalistic. In his 1906 *Harmonielehre* book he presented the analogous "equations" to explain the biological imperative that tones shared with living organisms. In nature, he explained, "creative urge → repetition → individual kind; In music: procreative urge

→ repetition → individual motif."[81] Just as it did in nature, repetition created form in music. Tones could be conceived of as creatures, with biological, procreative urges to generate individual motifs.[82] The motif then lived a fate, like a character in a drama.[83] If and when discovered, the motif was the only means of associating ideas between music and nature and therefore becoming art.[84] I highlight Schenker's ideas here to suggest that Kulke was not alone in attempting to employ biological concepts to the origin and development of music.

Returning to Mach and Kulke, both thought about evolutionary theory in similar ways. Both believed that humankind was subject to a Hering/Lamarckian inheritance of acquired traits, both physical and psychical. For Kulke, these heritable psychical traits included musical aesthetics. And Mach was evidently beginning to combine his understanding of sense organs and aesthetics within an evolutionary framework as well. Both Mach and Kulke understood the individual experience of listening to music to be psychophysical and evolutionary. This understanding was intertwined with their musical aesthetics. Because the aesthetics of music and the psychophysics of sound sensation were bound together for Mach and Kulke, it was possible for them to extrapolate from faith in their own subjective listening experiences to claims about the nature of hearing for humanity as a whole. Individual experiences of, say, accommodation or Wagnerian opera, were real and legitimate, dependent on the psychophysical apparatus of the individual and drawing on the values and aesthetics of the individual's time and place. Correspondingly, sound sensation in the human species as a whole was bound to time and place, potentially changing over time and between different locales.

From the Aesthetics of *Tannhäuser* to the Phenomenology of Psychophysics

I have claimed that Mach's search for the mechanism of accommodation in hearing drew on his and Kulke's psychophysical evolutionary theories, and, further, that these psychophysical evolutionary theories were intertwined with their musical aesthetics. This begs the question of what exactly these musical aesthetics were. We know that Kulke enjoyed Wagner, and from very early on. We can similarly assume that Mach, especially because he considered the cadre of Viennese Wagnerians lifelong friends, was at the very least, listening to the music of the New German School as well.[85]

The experience of listening to Wagner, however, changed during the second half of the nineteenth century. It no longer was the music or the musical experience that formalist listeners like, say, Helmholtz sought. Let us turn now to a brief discussion of Wagnerism generally as well as his early opera *Tannhäuser* specifically, to highlight some points of intersection between Wagner's conception of the *Volk*, Darwinian/Lamarckian evolutionary theory, and perhaps even the phenomenological implications of psychophysics.

In the 1850s, when Kulke was so struck by *Tannhäuser*, Wagner was better known for his writings in music criticism than his musical compositions. His jarring new harmonies, his use of leitmotifs, and his efforts to create the *Gesamtkunstwerk* (total work of art) garnered a devoted but small group of followers led by Franz Liszt. But in the following decade, Wagner's meddling in Ludwig's Bavarian court, his friendship with Friedrich Nietzsche, and his romance with Cosima von Bülow (Liszt's youngest daughter, married to Hans von Bülow) all increased his personal notoriety beyond the bounds of his music. As discussed previously, very few even respected Wagner's music. To his followers, however, Wagnerism was more than simply an appreciation for Wagner's compositions. It was a cultural, social, and political movement. This movement was fully realized by the breathlessly anticipated and highly celebrated 1876 opening of the Bayreuth Festspielhaus for the annual performance of Wagner's monumental cycle *Der Ring des Nibelungen* and *Parsifal*. The Bayreuth Festival quickly became—and remains to this day—a pilgrimage destination for Wagner admirers.

Wagner completed the libretto for *Tannhäuser und der Sängerkrieg auf Wartburg* (Tannhäuser and the Singers' Contest at Wartburg) in 1843 and the score in 1845. The opera was an examination of the struggle between profane and sacred love, Tannhäuser's choice between the temptations of the goddess Venus and the pious Elizabeth. The overture (the opening strains of which Kulke so loved) outlined the terms of the drama: a heart-weary Pilgrim's Chorus rises only to fall, interrupted by the tumescent, chromatic figures suggestive of the soul-consuming flames of wickedness, only to be (in the original Dresden version) overwhelmed by the thundering of the Pilgrims' hymn.

The figure of Tannhäuser himself, though perhaps a weak and manipulable hero (some critics have said that Wagner himself didn't understand

Tannhäuser, because he was such a weak character), promoted the value of individual aesthetic experience. His eventual choice of sacred love was all the more sacred and legitimate due to the fact that he had directly experienced profane love in Venusberg. Choosing sacred love with Elizabeth, Tannhäuser returned to a historical setting, the human world, insipid and frustrating as it may have been.

The opera was itself exceptional relative to many of Wagner's other operatic dramas in that the mythological world of Venusberg and the historical world of the *Volk* are separate. The conflicted Tannhäuser chose Elizabeth and the historical, human world. Wagner scholar Dieter Borchmeyer argues that central to Wagner's aesthetic ideology was his belief that "the musical drama was born out of the spirit of the Volk and that its creator drew directly on the figures of the mythopoeic popular imagination."[86] Wagner himself repeatedly claimed that the basic ideas behind his various works were from popular Volk tradition.[87]

I am hesitant about pushing this too far, but there are, I think, some elements of this effort of Wagner's to claim popular, Volkish origins for his musical figures and dramatic themes that are evolutionary, especially in the case of Tannhäuser's choice of a historical world clearly separated from the mythical Venusberg. We know from Cosima Wagner's diaries that Wagner was reading Darwin's *Origin of Species* and *Descent of Man* in the early 1870s. She described her husband's observations that Darwin was simply a further elaboration on Schopenhauer's ideas, in particular that all life strives to preserve itself and engender new life, and that instincts and mental faculties are tools to that end.

For Schopenhauer, the species were fixed and unchangeable but aesthetics were developmental in the sense that the aesthetic experiences of those with a high degree of genius were communicated to others through works of art. The art form of music was, for Schopenhauer, the purest manifestation of the striving Will. With this in mind, Wagner's efforts to locate the genealogical origins of his musical motifs in an older, Volkish aesthetic appear similar to Kulke's project of tracing Beethoven's Symphony in C Minor back to a traditional Bauerntanz. Not unlike the approach Riemann took in his music history studies, both Wagner and Kulke sought melodic origins that legitimized their present aesthetic priorities. Their respective efforts both included a (very) loose conception of Darwinian evolutionary theory.

Indeed, in his own, individual experience of music, Kulke drew on elements of Wagner's aesthetic ideology, and then combined them with the psychophysical and evolutionary conception of sound sensation that he shared with Mach. A quick example: Kulke was part of a conversation at the Bayreuth Festival in which he argued with a man who had declared the singing dragon of *Siegfried* to be nonsense pure and simple. Kulke used the exchange to explain to his readers that "not only is the dragon not anthropomorphized, but the tendency of the spirit of the scene is to represent man as a thing in nature, as a part of the whole in the great complexity of the All."[88]

At first blush this passage suggests that Kulke viewed Wagner's *Siegfried* through a Schopenhauerian lens: the work was about the individual's capability of merging with nature, potentially surrendering himself as an individual to, and of losing himself in, the great All.[89] But it is worth considering the extent to which Kulke's interpretation of the Wagnerian relation of humans to nature recalls the philosophical consequences of Mach's psychophysics: his increasingly phenomenological and positivist approach to the world. In *Die Analyse der Empfindungen*, Mach described the superfluity of the Kantian "thing in itself" abruptly dawning on him, stating, "On a bright summer day in the open air, the world with my ego suddenly appeared to me as *one* coherent mass of sensations."[90] This realization fueled his belief that one should approach the world as if it were made up entirely of elements of sensation—reality, appearances, one's self in the world—all one buzzing mass. A person did not surrender as an individual into nature, but rather was already part of it. When one conceived of the world as a single coherent mass of sensations, as Mach did, the individual was only one perspective within a complete psychophysical whole. Fechner had eliminated the causal connection between the physical and the psychical, replacing it with mathematical dependence only. With his development of psychophysical monism, Mach dispensed with Fechner's two-sided nature in favor of complete unification.

The iconic illustration of this unification of the psychical and physical into psychophysical monism, from Mach's *Die Analyse der Empfindungen*, can be seen in figure 4.9. The brow of the figure's eye and bridge of his nose frame the scene in which the figure is reclining in a room, writing or drawing something on a sheet of paper. The viewer only knows what the figure sees. What was being put on paper, say, knowledge of the material

Figure 4.9
Illustration of Mach's psychophysical monism, *Die Analyse der Empfindungen.*

world (both the human world of the room surrounding the figure and nature beyond, through the windows), was only accessible through the senses. *Die Analyse der Eimpfindungen* and this illustration appeared in 1886 but Mach had been thinking in these terms for more than a decade. And he had been sharing these thoughts with Kulke. Mach included an earlier sketch of this image in a letter he wrote to Kulke ten years prior (see figure 4.10). In their correspondence, Mach and Kulke were clearly discussing evolution, psychophysical monism, and of course, the conversation that brought them together in that café: musical aesthetics.

Conclusion

By 1885 the discussion of the accommodation mechanism of sound sensation had all but disappeared from Mach's writings. He never did find it.

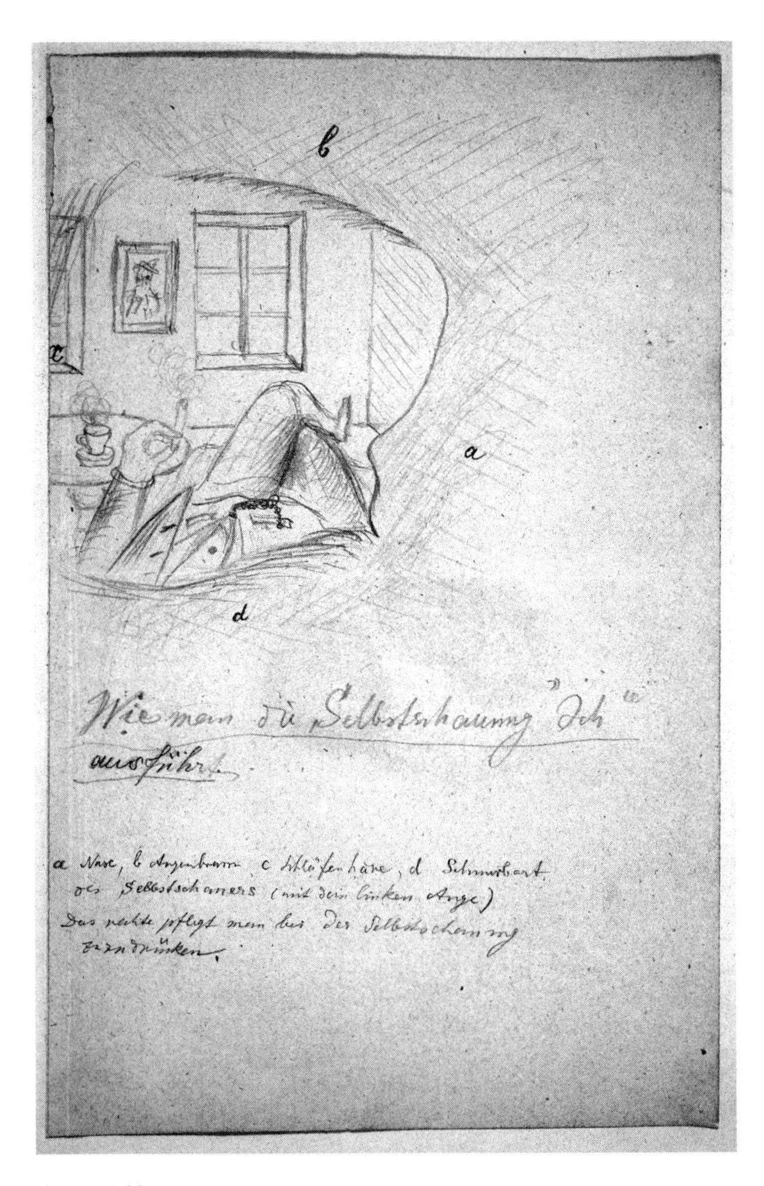

Figure 4.10
Ernst Mach, undated (between 1876 and 1878) letter to Eduard Kulke. Ernst Mach to Eduard Kulke, undated (though likely between 1876 and 1878) letter (no. 24), Ernst Mach Papers, Dibner Library of the History of Science and Technology, Smithsonian Institution Special Collections, Washington, DC.

But, between his psychophysical studies and his dialog with Kulke, Mach's historicist understanding of both sound sensation and musical aesthetics was fully formed. In *Die Analyse der Empfindungen* Mach asked how the development of modern music and the sudden appearance of great musical talent contributed to the survival of the species. Why did humans possess such fine discrimination of pitch, sense of intervals, or acoustic coloring that so far exceeded necessity or even usefulness? After all, according to Mach, music "satisfi[ed] no practical need and for the most part depict[ed] nothing."[91] He concluded that individuals developed their discriminating tone sensation—of pitch, of harmony—much like they developed their ability to distinguish lines, "as a sort of collateral product of [their] training, a sense for the agreeable combination of lines."[92] The ability to create and appreciate music, for Mach by 1885, was a byproduct of evolution:

> To deny the influence of pedigree on psychical dispositions would be as unreasonable as to reduce everything to it, as is done, whether from narrow-mindedness or dishonesty, by modern fanatics on the question of race. Surely people know from personal experience what rich psychical acquisitions they owe to their cultural environment, to the influence of long-vanished generations, and to their contemporaries. The factors of development do not suddenly become inoperative in post-embryonic life.[93]
>
> —Sound sensation was, not just for the individual, but for the human species as a whole, historically and culturally contingent.

I have aimed to push the origins of Mach's historicism back to 1863, the year in which he first met Kulke and Kulke's dilemma, and the year in which he first began his search for the accommodation mechanism of sound sensation. Much of what I have presented has been an analysis of the two men's published works highlighting points where psychophysical studies of sound sensation, musical aesthetics, and evolutionary theory intersect in order to locate the origins of Mach's psychophysical, historicist conception of sound sensation. If any doubt remains about the early date of 1863 it should be noted that Mach himself described his work since 1863 as "historico-physical investigations."[94]

This chapter began with the assertion that the opening strains of *Tannhäuser* structured Kulke's aesthetic heresy. By the end of his life Kulke had pushed his hope for a balance between objective analysis and subjective experience of music to the point of a battle cry. His call for the "equality of the sensations" (*Gleichberechtigung der Empfindungen*) in the face of the dogmatism of authority, not only reflected the unstable condition of musical aesthetics at the time but was also loaded with political meaning, clearly functioning as a critique of the political climate of Vienna at the end of the nineteenth century.[95]

Kulke's struggle to resolve his personal appreciation of Wagner with the dictates of Western music theory paralleled Mach's psychophysical search for the accommodation mechanism of the ear. Both men sought to reconcile the individual's experience of music with musical aesthetics. I have admittedly looped about and ranged rather far in this chapter. I have done so in order to highlight the points of interaction and overlap with the hope of demonstrating the way Kulke and Mach's parallel sound sensation projects, solutions, and aesthetics were products of the ways they moved through and engaged the priorities of the musical worlds. This informed the way they listened. It informed the way they thought about listening. The development of Mach's historicist conception of sound sensation must be regarded in relation to his friendship and intellectual exchange with Kulke. Their overlapping, reinforcing, fully integrated interest in the relation of the psychophysics of sound sensation to musical aesthetics motivated both men to integrate a distinct form of Hering/Lamarckian inheritance into their respective intellectual projects. The goals of their studies of the psychophysics of sound sensation and musical aesthetics were not mutually exclusive but instead reinforcing and fully integrated.

This union of psychophysics and the music world—as seen in the close friendship of Mach and Kulke—informed new ideas, very early on, about the historicist nature of knowledge. Because Mach attempted to understand individual musical aesthetics through experiments on the accommodation mechanism of hearing, he not only came to see that musical aesthetics—for both the individual and the human species—was historically and culturally contingent but that hearing itself was historically and culturally contingent. It was in his early psychophysical studies of sound sensation that Mach's historicism was first fully articulated.

5 The Bias of *Musikbewusstsein* When Listening in the Laboratory, on the City Streets, and in the Field

Not even the introspectionist himself has yet succeeded in maintaining clear vision with the eye rotated through 180° to see the mind that is at work. From this point of view we would seem to have a long way to go, and yet I must confess to you, attractive as my picture is, that I am not sure that we want to go, or can go, all that way.

—Edwin Boring, presidential address before the American Psychological Association, December 28, 1928[1]

Thus far this book as been an examination of how new musical aesthetics intertwined with new conceptions of sound and new conceptions of hearing. From the 1840s through the first decade of the twentieth century, the exploration of the sensory perception of sound was intimately related to a series of shifts that occurred in both science and music. The earliest forms of psychophysical study of sound sensation mobilized the unique measurement capabilities of psychophysics to make universal claims about the processes of hearing. By the end of the nineteenth century, in both music circles and psychophysical circles, the study and understanding of sound sensation was instead framed in terms of the subjective experience of the individual listener.

And as discussed in earlier chapters, in the first half of the nineteenth century the act of closely following musical performances was becoming one of the highest forms of aesthetic experience. It was also a social act in that it demonstrated membership in and reinforced the behavior and values of the upper-middle-class liberal elite. The act of unflagging attention was the only means by which to access the meaning of the newly ascendant autonomous music, the symphonic genre in particular. Attention itself had become an aesthetic experience.

By the end of the nineteenth century conceptions of sound sensation began to shift. Both in the psychophysical laboratory and on the city streets, discussions of hearing and listening became bound up with language about individual musical skill and musical training. The emphasis on the proper way to listen to music was fully aligned with the acceptance of subjective, individual listening in psychophysical studies of sound sensation. Refinement and musical training were required to do both. And in both music circles and psychophysical circles, the study and understanding of sound sensation were framed in terms of individual experience of the experimental subject. Further, according to some music critics, this advanced musical skill left its possessor vulnerable to unconscious bias from within and psychophysical assault from the outside world. The two examples examined in this chapter reveal that the status of musical expertise, both in the narrow confines of the laboratory and on the part of the music world's clamoring listening public, was renegotiated at the end of the nineteenth century. It was alternately valuable and menacing.

As an aesthetic experience, listening was inscribed with codes and values; there was a right and wrong way to do it. The skills valued in sound sensation were changing in the late nineteenth century, the consequence of upheaval in both the music and scientific worlds. So, at this very moment, when musical aesthetics and the psychophysics of sound sensation were the most intertwined, new priorities in the laboratory arose. To some, music was no longer equivalent to sound. Musical skills were not necessarily to be celebrated as scientific skills. The ability to hear interference beats or identify a pure fifth, for example, was increasingly considered an unimportant scientific skill. This chapter focuses on the parallel negotiation over the role and value of musical expertise in first the public space of concert halls and music criticism, then in the narrower confines of the psychology laboratory. Although there was no direct exchange between the work of Carl Stumpf or Wilhelm Wundt and the music critic Eduard Hanslick, there were interesting points of intersection that this chapter aims to highlight. Ultimately, we can understand the negotiation over the value of musical skill, both in the laboratory and on the city streets, to be part of a larger trend that replaced universal descriptions of hearing processes and highly social listening practices with a preoccupation with the subjective experience of the individual listener. These developments coincided with the rise of psychology as a laboratory-based discipline at the end of the nineteenth century.

The Changing Cultural Niche

In an 1885 article, the mathematician and philologist Alexander Ellis declared there was no single, "natural," universal musical scale.[2] Many musical scales of the world were not even based on laws, but were instead "artificial" and "capricious."[3] He based this conclusion on a detailed comparison of the musical scales (specifically, the number of tones and the types of intervals between these tones) of Greece, "Arabia," India, "Java," China, and Japan. He gained his knowledge of the non-European musical traditions from written treatises, what he could construct from instruments, and performances by native musicians visiting England. In a few instances Ellis attempted to relate the diverse musical traditions historically—for example, by asking if India drew on Arabia or the converse. Ellis's declaration did not lead to an immediate collapse of the belief in a universal musical aesthetic that just so happened to be Western. But it was taken seriously.[4] And it should be viewed as a clear indication of the direct effect the introduction of non-Western music—in part a consequence of the forced cultural exchange of colonialism—had on Europeans' conception of musical aesthetics by the mid-1880s.

Complementing the growing public interest in foreign music was an increased interest in local traditions, demonstrated by the rising popularity of folk music. These new musical traditions introduced alternative musical systems and destabilized the accepted structures and harmonies of Western art music. This destabilization can be seen in the increasingly atonal compositions of such composers as Richard Strauss, and later, Arnold Schoenberg and Alban Berg. The arrival of non-Western music, folk song, and atonal composition introduced new sounds to the German-speaking music world. In some instances these new sounds were welcomed with curiosity. Within a decade their reception was often an explosion of anger and violence. Concert riots like the one at the premiere of Igor Stravinsky's neoclassical opus *Le Sacre du Printemps* were not uncommon. The aesthetics of music were rapidly changing.

Musicologists and historians of music highlight Ellis's article, as well as the many native music performances at the 1889 Paris Exhibition, as significant sources of inspiration for European composers at the time.[5] The standard example is the music of French composers Claude Debussy and Maurice Ravel. Debussy's extensive use of pentatonic scales common to African and Indonesian music traditions was a consequence of his encounters with native musical performances at the 1889 Paris Exhibition.[6]

Ravel also, along with such composers as Béla Bartók, Leoš Janáček, Manuel de Falla, and Igor Stravinsky, embraced folk song and themes of preurban life in an attempt to mobilize their own *Volk* heritage against the perceived German and Austrian stylistic dominance. This impulse to employ folk material as a means of escaping the twin shadows of Beethoven and Wagner was both an effort to develop new sounds and part of a growing nationalist impulse. Bartók, for example, motivated by both the 1896 millennial celebration of Magyar presence in Hungary as well as folk nationalism generally, sought to collect and classify an expanded repertoire of Hungarian folk tunes. This project fueled his efforts to develop a Hungarian modernist art music. Musicologist Julie Brown has traced the shifts in Bartók's Magyar nationalism from these narrow early efforts to include an increasingly wide range of ethnic music, buttressed by changing conceptions of hybridity.[7]

Complementing the nationalist folk music impulse was an effort to move past the Wagnerian tradition—Wagnerian music had by this time become establishment music—through the deliberate rejection of the Western tonal system. In developing his twelve-tone approach and, later, atonal music, Arnold Schoenberg drew on the work of Helmholtz to actively create dissonant intervals.[8] He described his work as an "emancipation of the dissonance" in that his style renounced a tonal center in order to treat dissonances like consonances.[9] Tonality, according to Schoenberg, was dead. His disciple Anton Webern later elaborated in a private lecture in Vienna, adding personal responsibility to the death of tonality, claiming that in their efforts to preserve the keynote and save tonality, "we broke its neck!"[10]

The Wundt-Stumpf Debate

It was this world of tonal upheaval, destruction, and retreat into which Wilhelm Wundt and Carl Stumpf were drawn in a fierce debate over the nature of tone differentiation. Given that one of the main elements of the controversy was the role of musical expertise in the laboratory study of tone differentiation, the instability of the music world must be integrated into an analysis of the Wundt-Stumpf debate. Several historical works have addressed specific aspects of the German psychophysical impulse, but they consider the role of music incidentally if at all. Historians have focused on Wilhelm Wundt's conceptual devices such as introspection and apperception, or his experimental methods, such as Kurt Danziger's work on his "methodolatry."[11] They have failed to devote any analysis to Wundt's role in the narrative I examine here, though many have noted his debate with Stumpf. Scholars of Stumpf, however, rarely mention this debate. I think this is due in part to the bifurcated nature of the Stumpf historiography. Several historical studies have examined Stumpf's work but have focused either on his ethnomusicological research or his experimental psychology and the Gestalt School he established.[12] This chapter does not necessarily aim to overcome this bifurcation and present Stumpf as a complete intellectual whole. Instead, an additional goal of the chapter is to illuminate an episode at the intersection of the many strands of Stumpf's career, where psychophysics and music met and intertwined.

Let us turn now to the fierce debate in the early 1890s between Carl Stumpf and Wilhelm Wundt. The controversy began with an 1890 publication by Wundt's student, Carl Lorenz, on a series of tone differentiation experiments he had done in the previous decade (Wundt had included some of Lorenz's results in the third edition of his *Physiologische Psychologie* in 1887). Stumpf and Wundt then each published follow-up articles. The final round of the exchange—featuring Stumpf's paper, titled "Mein Schlusswort gegen Wundt," versus Wundt's paper, "Auch ein Schlusswort" —was dominated by attacks and jabs at the other's argumentation techniques and conduct. In the end, the debate was no longer about the experimental challenges of measuring tone differentiation. In part this was because, as Edwin Boring explained in his 1928 presidential address before the American Psychological Association, while Stumpf started the controversy, Wundt had made it personal.[13]

Boring had presented the Wundt-Stumpf debate, parsing out the subtext of the increasingly personal attacks, as an example of one of the paradoxes of human nature that limit the work of scientists: the more one fought for the truth, the less he or she saw it.[14] The polemic was initially framed in terms of the proper application of the Fechner-Weber law to studies of pitch sensitivity. In question was whether pitch differentiation —distinguishing between two very similar tones—was analogous to just-noticeable difference as Fechner and Weber had originally conceived of it.

Though the dialog initially began as an argument over the (mis)application of the Fechner-Weber law, the debate soon turned to the issue of experimental design and observer expertise. Stumpf and Wundt squabbled over the importance of the experimental subject's musical expertise and the nature of the scientist-as-expert's role—refined listening skills or rigorous self-observation and analytical skills. Further, as Boring's subsequent analysis demonstrated, this clash over the proper experimental approach and interpretation of data was infused with personal attacks on the other's intelligence and integrity.[15] Ultimately, one of the issues at stake was the role of musical expertise in sound sensation studies and the bias of Stumpf's conception of music consciousness, *Musikbewusstsein*.

Stumpf was originally trained as a philosopher; having looked to Gustav Fechner for guidance, Stumpf decided in the early 1870s to devote himself to "that field which, connecting my musical experiences and studies with the interests of psychology, seemed to me, personally, the most promising."[16] He took up work with acoustic devices and worked closely with the instrument maker and acoustician Georg August Ignatius Appunn, who had also competed with Rudolph Koenig to develop instruments for Hermann Helmholtz and then, in 1875, turned to the psychological aspects of hearing. In his 1883 two-volume treatise, *Tonpsychologie*, Stumpf introduced his psychological theory of tone fusion as a more sufficient explanation of sound sensation than Helmholtz's strictly physical approach.

Wundt was similarly an admirer of Fechner, and as with Stumpf, his studies of sensory perception were a departure from Helmholtz's foundational work on sound sensation. Wundt had trained in physiology under Johannes Müller and Emil Du Bois–Reymond at the University of Berlin, and had been an assistant to Helmholtz in his physiology laboratory at Heidelberg from 1858 until 1865. His turn toward experimental psychology in the 1860s reflected changes occurring within the discipline of physiology, which, in large part due to Wundt's efforts, had recently been reconceived as an experimental science. In 1879, following through on directives included in his 1874 *Grundzüge der physiologische Psychologie*, Wundt established the psychology laboratory at Leipzig, the first of its kind, dedicated to introducing experimental methods to psychology complete with the journal *Philosophische Studien* to publish the methods and results of the laboratory's research. In the early decades of the Leipzig lab (the period relevant to us), this research was predominantly devoted to examining mental process duration, the sense of time, attention and memory in relation to the association of ideas, and the analysis and measurement of sensation.

In pursuit of this last line of research, Wundt's student, Carl Lorenz, presented in an 1890 article the results of his experiments on *Tondistanzen*, the ability to judge the middle tone between two tones sounded in series. These *Tondistanzen* experiments were essentially measurements of just-noticeable differences in pitch much like those performed by Ernst Heinrich Weber fifty years earlier. Observers listened to a lengthy series of tones, increasing in increments of four cycles per second. The first and last tones (termed *Grenztöne*) were either musical or nonmusical intervals. Lorenz asked the observers to indicate when the middle tone between the Grenztöne was reached. He sounded the series of tones both from low Grenzton to high Grenzton and from high to low, recording more than 110,000 judgments.[17] In part, the Leipzig laboratory's series of experiments demonstrated that the method of just-noticeable-difference measurements, at least in the way they applied them, could be successfully employed to examine tone sensation.

Further, the results showed that the Fechner-Weber law did not measure absolutes but instead proportionalities in the case of sensing tone difference. Lorenz's data analysis suggested that observers tended to choose the arithmetic mean as the middle tone. That is, if A and B are the Grenztöne and C is the middle tone, then $C = (A+B)/2$. In Western tuning systems (both tempered and nontempered), the arithmetic mean would only be a musical interval when A and B were two octaves apart. So Lorenz's analysis showed that observed tone distances were likely unrelated to musical intervals.

Lorenz used the results of his experiments to critique the tone fusion work done by Stumpf in the previous decade in *Tonpsychologie*. Stumpf had shown that tone fusion (the perception of two simultaneous tones as one single tone) occurred most frequently for musical intervals. Lorenz elaborated that Stumpf's great error was his use of musically trained experimental subjects. Because these individuals were more skilled at identifying certain intervals of tones than in following instructions for psychological experiments, Lorenz suggested, they biased Stumpf's data.

Stumpf replied with a scathing and lengthy article that meticulously picked apart Lorenz's study, from the format of his tables to his treatment of experimental error.[18] He also rearranged Lorenz's measurements from a "musical point of view" (*musikalischen Gesichtspunkten*) in order to reveal what he believed to be a significant and systematic oversight in the study. For the Grenztöne interval of a single octave, where the judgments of the middle tone were significantly more definite (that is, with more judgments of the middle than of the surrounding tones, so, a small standard deviation in the distribution of judgments), the middle tone corresponded to the geometric mean or "musical" middle. The geometric mean, C, is determined by the equal ratios (again, when A and B are the Grenztöne) of $C/A = B/C$ or $C = (AB)^{1/2}$. So for example, when A and B form an octave (c-c′), C is the interval of a fifth (c-g). Stumpf found this tendency of observers to choose the geometric mean as the middle tone to be the case for other musical intervals as well.

Accordingly, Stumpf found the judgments of the middle tone were more definite when the middle tone corresponded to the "musical" middle. This correspondence occurred again and again, indicating, Stumpf claimed, a critical role for musical abilities.[19] Sensed tone distances *were* related to musical intervals. The ear heard musical intervals out of musical habit, he argued. An individual's hearing was subject to what Stumpf termed music consciousness (*Musikbewusstsein*).[20]

Stumpf was also troubled by Lorenz's apparent conflation of subjective and objective, or rather, the assumption that subjective judgments coincided with the objective reality of sound-wave frequency.[21] He felt that if judgments were determined to be subjective, they should not be described in terms of frequency. What did a comparison between the middle tone, the arithmetic middle, and perhaps the geometric or musical middle say about the truth or reality of what the individual sensed—the sensed middle?[22] An understanding of the difference between tone ratio and tone distance, the sensed middle Wundt and Lorenz sought, would be achieved only by paralyzing the musical habits of the observers. But this would be an extreme abstraction from the usual conditions of the phenomenon of tone sensation.[23] No law could be developed from this information. Wundt and Lorenz's work was, according to Stumpf, so arbitrary that they might as well have "put it to a vote of the people."[24]

Stumpf's reinterpretation of Lorenz's data brought him to the following conclusions: First, with the exception of the double octave Grenztöne set, the observed/sensed middle tone was the musical middle rather than the absolute middle. Second, when the musical middle was unclear the judgments varied greatly and were significantly less definite. Third, for nonmusical intervals, the observed/sensed middle tone was heard as the nearest musical middle, though admittedly with significant variation in judgment. Finally, the musically gifted and skilled observer made the most definitive judgments. These results were only comprehensible, Stumpf argued, when musical-interval consciousness (*musikalische Intervallbewusstsein*) was accepted as the most influential factor.[25]

In his defense of Lorenz's work, Wundt countered that Stumpf's criticisms were misplaced and rooted in a misunderstanding of Lorenz's research program. Further, Stumpf did not understand how to properly apply psychophysics. Wundt continued, "Just because one cannot reach the Moon, does not mean that the distance to it from the Earth cannot be measured." He asserted that Stumpf was, in fact, undermining psychophysics.[26]

Stumpf had, of course, previously complained that Lorenz's research program undermined the "general law of relations" (*allgemeinen Gesetzes der Beziehung*) of which the Fechner-Weber law was a special case. This general law applied to the entire psychical field. Stumpf lamented that by exposing the contradiction within the Fechner-Weber law as applied to tone sensation, Wundt's students rendered the general law void.[27] Notice that in this opening round of the debate, for both Wundt and Stumpf, undermining the Fechner-Weber law was sacrilege and an accusation each launched at the other.

Although both parties accused the other of misapplying the Fechner-Weber law, both also claimed that the law did not apply to all areas of tone sensation. Some of the vitriol of the debate may have been rooted in anxiety about the law's continued relevance to psychophysical experimentation. Lorenz believed that psychophysical experimentation had established that the Fechner-Weber law was invalid for both small and large pitch differences, and although Stumpf found strong support for the law for pitch differences that corresponded to Western music intervals, he did not focus on that point. He instead emphasized the implications of the law for psychophysical studies of sound sensation, and the consequence that experimenters must employ musically skilled observers exclusively. Stumpf had stated outright that he did not believe the Fechner-Weber law to be "the Alpha and Omega of all sense-psychology research."[28] He reiterated this position again in his response to Wundt, explaining that the broad field of tone psychology could not be united through one cohesive theory.[29]

Following this first exchange, the debate turned away from the proper application of the Fechner-Weber law to, among others, the issue of observer expertise and experimental design. Stumpf insisted that judgments of tone distances by musically trained observers were different from those without musical training, and that they were better. Only musically skilled observers were capable of judging middle tones, of comparing pitches. Thus, Stumpf continued, Lorenz and Wundt should have used only musically skilled observers.[30] Some of Lorenz's unskilled observers apparently could not even discern which of two pitches was higher.[31] According to Stumpf, "a single judgment by one such [musically skilled] observer [was] worth more than a thousand by the unmusical and unskilled."[32]

Stumpf's repeated insistence on the musical skill of observers in sound sensation studies and his need to explain and defend the relevance of musical expertise reflected an anxiety about the decreasing importance of musical expertise in the laboratory. Certainly Stumpf was suspicious that Wundt's theories were divorced from the reality of sound sensation. What bothered Stumpf about Wundt's discussion was twofold: First, Wundt seemed to make a strange and contradictory lack of distinction between sense and idea, using "Empfindung der Klangeinheit" and "Vorstellung der Klangeinheit" interchangeably. Stumpf admitted that this distinction did not so much matter as seem arbitrary since they were one and the same psychical activities. Wundt was making a distinction that Stumpf did not think existed.

Stumpf's second issue was with Wundt's proposal that for the six-part chord of tones related by the ratio of 1: 2: 3: 4: 5: 6, when just the higher three tones are sounded, the idea of one harmonic of several tones would appear while the same sense of a tone singularity would not be felt. With the addition of the three lower tones, however, the idea of one harmonic of several tones immediately ceased, leaving only one single sound. Stumpf stated that he was totally unconvinced of this effect.[33]

Stumpf believed Wundt's conclusions to be based on highly questionable observations. He did not trust Wundt's hearing, especially his musical skills.[34] Also, Stumpf was troubled by Wundt's discussion of listeners' ability to hear interference beats. Wundt disputed whether a listener could detect a frequency of interference beats greater than 60 beats per second.[35] This was the upper limit that Wundt's "expert" (*Kenner*) was capable of detecting.[36] Stumpf pointed out that this frequency was about half that which Helmholtz claimed to be able to detect (up to 132 beats per second), and that he himself could detect even higher beat frequencies (264 beats per second!). Stumpf questioned not just Wundt's ability to distinguish discrete musical tones but his ability to hear at all.

In his rebuttal, Wundt objected to Stumpf's declaration that only musically skilled observers should be employed in sound sensation studies. Wundt believed that the judgments of musically unskilled observers were perfectly usable. In fact, these observers' *lack* of previous musical experience made their observations "particularly estimable" (*besonders schätzbar*).[37] By excluding musically unskilled observers, Wundt explained, Stumpf lost valuable data. For Wundt, the value of data lay in its volume; the more observations the better. This is not to say that there wasn't a role for thoughtful and controlled subjective judgments by experimental observers for Wundt. Indeed subjective testimony was at the core of the introspection technique of self-observation he developed and promoted. Rather, Wundt was concerned that musical training would bias the observers' introspection. He even considered passive, not-deliberate hearing to be a form of bias. So this was not a distinction between (active) listening and (passive) hearing.

Instead, experimental expertise for Wundt lay in both the proper use of introspection and in the ability to properly analyze large volumes of data, not in the musical skills of the experimental subject. Further, by Wundt's analysis, there was no indication of a musical skill-dependent bias. The fundamental levels of consciousness could not be music-infected. It would only appear so, claimed Wundt, if one's analysis was limited to a superficial inspection.[38] He continued, "if Stumpf has an insurmountable dislike of arithmetic, that is his problem; nobody is forcing him to do psychophysical research."[39]

If Wundt saw value in large aggregates of data, Stumpf did not. For him, to include data from unmusical observers, and equate it with that collected from the musically trained, was to ignore the differences between observers. Such an act distorted the data to the point that it was completely arbitrary and might as well be voted on by the masses rather than executed by natural scientists.[40] He equated musical skill with scientific skill. The democratization of observation, the move away from the (musically trained) individual, was troubling for Stumpf. His remarks on Wundt and Lorenz's experiments suggest a concern that the masses of mediocre observers and their flood of data points threatened to overwhelm the uniqueness of the individual, musically trained expert, and his personal contribution to the psychological understanding of acoustic and musical phenomena.

The debate ended anticlimactically. Later in his autobiography, Stumpf claimed he had been stung by the sharpness of Wundt's invective. He also claimed that he was objectively correct and that this was proven by the fact that the results of the experiments of Wundt's laboratory, supposedly revolutionary enough to have overturned the Fechner-Weber law, were never mentioned again anywhere, except in Wundt's textbook.[41]

The debate's turn from maintaining the integrity of the Fechner-Weber law to psychophysical expertise reflects a significant shift in the history of psychophysical studies of sound sensation. Whereas previously the goal of psychophysical studies of sound sensation had simply been to develop theories of hearing based on measurable features, there arose, in part due to changing goals within the developing field of experimental psychology, a new goal of adjudicating the right and wrong way of hearing. Musical skill was losing value and validity in the laboratory. So, as standards for measurement and the interpretation of measurement were negotiated and narrowed, we observe a case of the experimental protocol tail beginning to wag the epistemological dog.

The debate ended in a draw. Boring later recounted that he could not, for the life of him, determine who had won.[42] Further, he explained that Stumpf could not have convinced Wundt to adopt his position, nor could Wundt have convinced Stumpf to accept the validity of his. Boring also noted that Edward Titchener, a British psychologist who studied with Wundt, described reading through the debate (remember, all of this was published) three times. Twice he sided with Wundt and once with Stumpf. In short, there was a lot of hand-wringing in the aftermath of the Wundt-Stumpf debate, partly because it was unseemly but perhaps also because it exposed the possibility that even the most rigorous engagement over a problem in experimental psychology could fail to produce a correct answer.

The Bias of *Musikbewusstsein*

Stumpf's conception of music consciousness, or "music-infected consciousness" (*musik-infinizierten Bewusstsein*) as he sometimes termed it, can be traced to his discussion of conditions that contribute to tone judgments in his work on tone fusion, originally presented in the first volume of *Tonpsychologie* in 1883. There he argued that a strictly physical approach to sound sensation (Helmholtz's) was insufficient and instead proposed that consciousness (as opposed to unconscious, purely organic conditions), and, ultimately, music-infected consciousness, was the greatest obstacle for just-noticeable-difference judgments of tone.[43] The terminology invoked during his clash with Wundt reflected Stumpf's ambivalence about the origins of music: the language of "music-*infected* consciousness" is telling. Although he implied that aesthetics was a disease, a foreign agent, he did not necessarily consider it a menace. If anything, the bias of Musikbewusstsein allowed for better insight into sound sensation. The implication of Stumpf's claims that observers tended to hear geometric means (musical middles), especially when the observers were musical, suggested that the applicability of a law determining tone distance sensitivity varied depending on the musical expertise of the observer. Perhaps the Fechner-Weber law only applied to the musically trained.

The bias of music consciousness was the product of specific musical training. This fact questioned the innateness of the Western tonal system, which Stumpf did address (and I will discuss shortly). Just as a performer could train his or her body to create certain sounds, Stumpf's conception of Musikbewusstsein suggested that listeners could train their bodies to receive certain sounds as well. Further, his language implied that a musically trained individual *could not help* but receive certain sounds. Though an individual's experience of sound was indeed individual and subjective, he or she could not exercise any control over this experience (as opposed to accommodation in hearing in which the individual's change in focus or attention resulted in a changed aural experience).

For Wundt, on the other hand, Musikbewusstsein contributed nothing to a psychophysical study of sound sensation. He instead privileged large aggregates of data rather than the subjective testimony of a few well-trained observers. Wundt's effort to deemphasize the role of musical skill in deference to experimental protocol, statistical analysis in particular, was a departure from the earlier psychophysical studies of sound sensation. Stumpf's reaction, which was to employ musical expertise as a criterion for credibility and as a means of dominating scientific debates, is additionally significant. The previous generation of sound sensation researchers had employed music and their own musical expertise in an unselfconscious way. Indeed the entire discussion of the relevance of musical expertise to experimental studies of sound sensation was new. Previously the use of music as an instrument of psychophysical investigation or means of argumentation had been completely uncontroversial; now the subjective, individual experience of sound—including music—was increasingly insignificant for studies of sound sensation in the Leipzig laboratory.

The increasing desire on the part of Wundt and his followers to eliminate the role of musical skill in the psychology laboratory left little room for individual musical aesthetics in the psychophysical study of sound sensation. Previously, because musical skill was considered a scientific skill, connoting superior ability to experience sound, the subjective, individual experience could stand in as the universal experimental subject. However, once the testimony of musically *un*trained observers was accepted as more legitimate and more universal, musical skill and therefore individual musical aesthetics became irrelevant. The conception of sound sensation, at least in a laboratory context, was increasingly defined in terms of what was experimentally testable by the statistical method.

A Musikbewusstein epilogue: in 1907, five months before Joseph Joachim's death, Stumpf wrote to the violinist inquiring about his hearing affliction.[44] Apparently Joachim was experiencing *Doppelhören*, an aural echo of some sort, likely caused by an infection in the middle ear. Stumpf requested that Joachim fill out a seven-part questionnaire (*Fragenschema*) about when and what he heard. He asked if Joachim could determine the tone. And he asked if the echoed tone in the sick ear was ever as clear and strong as the correct tone in the healthy ear so as to create a perfect feeling of a double tone (*Zwei-Klanges*). This of course cannot be pushed terribly far, but I very much like that Stumpf was asking one of the greatest musicians of the nineteenth century to make an observation of his tone differentiation during an ear infection; *musik-infinizierten Bewusstsein*, indeed.

A New Aesthetics of Listening

From the 1880s on, Stumpf published a series of studies of non-Western musical traditions based on performances that he witnessed live and through recordings made on Edison cylinders.[45] In this way, Stumpf's work moved out of the traditional confines of the laboratory. Indeed, in his 1911 work, *Die Anfänge der Musik* (The Beginnings of Music), Stumpf suggested that humans were capable of judging pure pitch distances as long as they weren't conditioned by the principle of consonance or hindered by the "false experiments" of the psychology laboratory. He claimed that if his analysis of the field recordings differed from the claims of the (Wundt) Leipzig laboratory, it was further "confirmation of the experimental errors therein."[46] Stumpf's study of sound sensation now consisted of observing recordings of non-Western peoples in the field almost exclusively.

In addition to the expansion of the sound sensation laboratory—as seen in the ethnomusicology fieldwork—there were also relevant developments in the work of musicologists and music critics during this period. Let us return briefly to the writings of Eduard Hanslick. Recall the many typologies of listeners he presented in *Vom Musikalisch-Schönen,* of which the aesthetic listener was the most refined and superior. The aesthetic listener, through the careful application of his or her attention to the sonically moving forms, would gain access to the beautiful in music. Such aesthetic listening, however, also left the listener vulnerable to aural assault.

In 1884, thirty years after *Vom Musikalisch-Schönen* appeared, Hanslick published the article "A Letter on the 'Piano Pestilence'" ("Ein Brief über die 'Clavierseuche'"), in which he decried the scourge of the modern city: the piano playing of neighbors.[47] The errors and missteps of "tinkling dilettantes" and "practicing students" assaulted the poor, defenseless ear of the city dweller. Hanslick lamented the anxiety of awaiting the well-known but inevitably misexecuted chord of the neighboring Fräulein and the permanent dulling of the ear perpetrated by the daily cacophony of the "piano pestilence." The torments to the refined listener, according to Hanslick, were both psychological and physical.

The language employed by Hanslick is provocative, and implies that musicianship was a disease (or at least a preexisting condition) that left its victims susceptible to torments fueled by the growing popularity of the pianoforte. Musical expertise manipulated one's aesthetic experience, not only in the concert hall but also on the streets of the cityscape. Again, like Stumpf's musically skilled observers, the discerning listener was doomed and defenseless against the piano epidemic. He explained that the only possible means of preventing further psychologically nervous, physiologically dulled victims of the "piano pestilence" would be a reduction in piano students: a sacrifice of the pathological performers for the sake of the aesthetic listeners.

Hanslick's formalist aesthetic was both a vehicle for his music criticism and a means of discussing the act of listening. Not only is the pseudo-evolutionary and medicalized language interesting, but the very fact that he introduced the expert listener into his discussion of musical aesthetics is noteworthy and new. In contrast to the concerns of such cultural critics as Max Nordau who dwelled on immanent physical, psychological, and, ultimately, cultural degeneration as "mystics" and "egomaniacs" turned away from hard work and discipline to produce retrogressive art and literature, or Friedrich Nietzsche, who observed modern society to be the stifled product of a past capitulation of the strong to the weak, Hanslick focused on the refined listener as the exception to the growing masses engaged in mediocre piano playing.[48] And yet the physical and psychological dangers of poorly performed music to the ears of Hanslick's listening elites were worthy of his readers' concern.

Hanslick's slightly sarcastic worries can be nestled into a much longer history of anxieties over the dangers of music to the health of both listeners and performers.[49] Musicologist James Kennaway has shown that the concept of music as physical threat to the nerves had existed since the end of the eighteenth century, its origins bound to the new material understandings of sound developed by physicists, physiologists, and acousticians. One might also recall Friedrich Nietzsche's piece, *The Case of Wagner*, indicting Wagner's music for the degeneration of both the social and the spiritual, causing both physical and psychical decay.[50] For Hanslick, however, it was not just music nor poorly performed music that caused suffering. It was also the listener's rigorous musical training that made him or her vulnerable—an unfortunate side effect of musical expertise. Hanslick's description of the persecution of the musically trained listener by the tedious clatter of piano students throughout the city indicates a shift in the relationship between composer, performer, and listener. The listener was not only relevant to the musical aesthetic project but, by the end of the nineteenth century, had to be catered to. The listener's, let us say, Musikbewusstsein needed to be protected and preserved.

This effort to control and/or preserve musically trained listening, seen both in the laboratory and on the city streets, can be viewed as part of a larger trend in the sciences to discipline the experimenter's body. Historians of this trend have focused on experimenters' "ways of seeing" in the field and the laboratory and have framed these ways of seeing as forms of tacit and embodied knowledge irreducible to formal and symbolic terms.[51] Lately scholars have expanded this notion of embodied knowledge to include the material culture of science—the objects and artifacts with which scientists interact—as well as the role of hearing and listening in the laboratory.[52] Musicologist Ben Steege, for example, discusses the development of autolaryngoscopy, the self-observation of the throat while intoning vowels that became a very popular pedagogical and experimental tool in the mid-nineteenth century. Laryngoscopy required the cultivation of both a bodily skill, which Steege terms "laryngoscopic perception," and a "punctual rupture in self-knowledge" that functioned as if the self-observing individuals had come into contact with themselves for the first time.[53] Historian Cyrus Mody, for another example, examines the techniques experimenters developed to accommodate auditory concerns through an awareness of their sound environment, supplementing their experimental practice.[54] He shows how sound was an additional and critically important part of tacit knowledge in the laboratory. The debate between Stumpf and Wundt discussed here illustrates not only how important specific forms of tacit listening knowledge were to the psychological laboratory but how contested this knowledge had become.

Throughout the nineteenth century, music and laboratory training required a refined ability to both generate and receive sounds. This refined ability was rooted in a specific aesthetic. The generated and received sounds, when they were musical sounds, were musical sounds of the Western tonal system only. Both the professionalization of experimental psychology and the introduction of non-Western music to the ears of Central Europeans at the end of the nineteenth century created new and different demands of musical expertise. The clash between Stumpf and Wundt over the role of Musikbewusstsein in tone differentiation studies as well as Hanslick's call to protect the tender ears of the musically trained from the cacophony of mediocre music reflect an anxiety over these shifting expectations.

The New Laboratory

Complementing these shifts in and associated anxiety about the status and form of musical expertise was a changing treatment of Western music's others. Some of these others were external, from beyond the traditional boundary waters of Europe, as mentioned at the beginning of this chapter. Some were internal, in the form of non-Christians. Musicologist Philip Bohlman extends the work of Edward Said to show how Europe's obsession with musical others reflects a consistent effort to "employ music to imagine its selfness."[55] The folk song, Bohlman continues, became an object "onto which Europeanness and internal otherness were juxtaposed" and gave rise to a competitive conflict between selfness and otherness.[56]

Bohlman directs us to Vienna, the center of the decaying Austro-Hungarian Empire, as the site of much of this negotiation at the end of the nineteenth century. It is worth noting that nearly all the music critics, musicologists, and ethnomusicologists covered in this work—Kulke, Hanslick, Schenker, and Hornbostel (to be discussed shortly)—were Viennese and were Jewish; specific examples of the internal others interested in external ones that Bohlman refers to generally. There has already been extensive historical study of Jews in Germany and the Hapsburg Empire, and much more can be said about the role of identity construction of internal others in intellectual and cultural history.[57] Acknowledging this, I have focused on the efforts of these individuals to define a Western musical aesthetic through a *wissenschaftlich* framework. At the end of the period examined in this book, these frameworks were directed at external others. This can be seen, in Stumpf's case, in his ethnomusicological turn and, as a consequence, expanded conception of the laboratory.

Stumpf ended his overt criticism of the Leipzig tone differentiation studies in 1892 and turned his energies to establishing his Berlin Psychological Institute and seminar (granted in 1894, in part to compete with Wundt's Leipzig laboratory).[58] The Berlin Psychological Institute, and the students Stumpf trained there, would later become the epicenter of the Gestalt school of psychology. In 1900 he further developed his already extensive collection of acoustical instruments to establish the Berlin Phonogramm-Archiv.[59] Under the management of his student Erich Moritz von Hornbostel, the Phonogramm-Archiv collected and archived field recordings of music from all over the world. So Stumpf did not so much end laboratory-based tone sensation studies as alter the boundaries of his laboratory. The work of the Phonogramm-Archiv was an extension of the self-observation and explicitly interdisciplinary practices the Berlin Institute had maintained since its inception.[60]

The self-observation Stumpf employed in the research for *Tonpsychologie,* the technique he also trained his Psychological Institute students in, and the expert practices he clashed with Wundt over, could be applied to the analysis of the field recordings.[61] Certainly the musical training would be considered absolutely necessary. The field recordings, collected at sites far beyond the frontiers of Western musical aesthetics, allowed for self-observation, Musikbewusstsein and all, to also extend far beyond the walls of the Berlin Psychological Institute.

The laboratory expanded physically as well. Stumpf's own move into the fledgling discipline of ethnomusicology can be seen as an indication of the legitimacy that the field now held as a site of serious and legitimate study of sound sensation. The phonograph had allowed sound to transcend time and space. A live musical performance was made mobile. And repeatable. The Phonogramm-Archiv was devoted to the comparative musicological study of the melodies of "less cultured" folk. Stumpf, the Berlin physician Otto Abraham, and Hornbostel had established a practice of providing recording equipment to anthropologists and ethnomusicologists in exchange for their original Edison cylinders (copies were then made and provided to the field scholars).[62]

Certainly the introduction of sound recording technology is pertinent to this narrative. Stumpf's use of cylinder recordings from the field, for example, meant that the experience of listening to the musical performance was often completely separated both temporally and spatially from the performance itself. Recall that when Helmholtz both generated and received the sound object, his own body was, ostensibly, the musical instrument. By the early twentieth century, the generation and reception of the sound object were no longer bound together. So perhaps the object of investigation was not exactly destroyed so much as reconceived in a material form, as an object in the traditional sense. The wax roll was transportable and preservable. Individual, subjective music making was no longer necessary for scientific knowledge making.

Compared to, say, Riemann's advocacy for active listening rooted in musical training in the application of his system of harmonic dualism, listening to a phonograph was active in a different way. The researcher could replay the recording as many times and at as many speeds as desired. In this way the researcher was liberated from the need to retain the subtleties of a performed piece long enough to hand-transcribe it to paper. Hornbostel further argued that, by circumventing this step, the researcher avoided slipping into European aesthetic assumptions. He did not use the term *Musikbewusstsein* but he did refer to the great effort to overcome powerful harmonies that could be avoided with the new physical-acoustic methods of comparative musicology.[63]

Additionally, practitioners believed that the reobjectification made possible by the phonograph granted the fledgling field of ethnomusicology a previously lacking scientific authority. As an instrument of observation, the phonograph was considered superior to the naked human ear. For example, during his collection efforts in the isolated regions of Central and Eastern Europe in the first decade of the twentieth century, Béla Bartók both transcribed the songs he heard by hand and used the phonograph, but he repeatedly articulated a preference for the phonograph or gramophone over a straight transcription into musical notation. The phonograph captured the gliding effects of the local music for which there was

Figure 5.1
Bartók recording folk song

no appropriate musical sign. Nor were there exact signs for tone color. Mechanical recording devices were, Bartók explained, indispensably efficient in the collection of music. More songs could be collected and the performers did not tire or become bored, as in the case of direct song transcription.[64]

Perhaps the most appealing function of the phonograph for Bartók was its ability to eliminate the subjective element of folk-song collection. While admitting that such an ideal might never be achieved, he expressed hope that the entire task of recording and classifying folk songs would eventually be entrusted to a machine.[65] Further, once a recording was made, the rotational speed of the cylinder or record could be slowed to allow for a listener to observe the melody "like an object under the microscope."[66] In this way the listener experienced "the very intricate or hardly perceptible ornaments and rhythmic differences with an accuracy unobtainable from hearing the natural song."[67] Like the microscope, the phonograph supplemented human sensory organs.[68] It was a scientific instrument.

For Bartók, this ability of the phonograph to objectively record and then amplify the folk song for analysis transformed folk-song collecting into scientific research. The measuring and fixing instruments allowed for the most faithful reproductions of the melodies (*snapshot* is the term Bartók used). The best equipment, however, was not enough. He explained that equivalent intellectual equipment in the form of extensive knowledge of linguistics, phonetics, dance choreography, and understanding of the relationships between music and customs, sociology, and history, was also required. Above all, Bartók continued, it was indispensable that the folklorist "be an observant musician with a good ear."[69]

I would like to argue that, especially in the case of the Phonogramm-Archiv, where the original cylinder recordings were maintained in Berlin, while data was collected from all over the world, the expert stayed put. As such, the walls of the laboratory did not simply fall down. They moved outward. An expert trained at, say, the Berlin Psychological Institute would be skilled in both self-observation and music. As a result, though the walls of the laboratory changed, the practices remained the same. The psychophysical study of sound sensation remained an aesthetic project.

Musical training therefore remained critical for a *wissenschaftlich* ethnomusicology, even as the boundaries of Western musical aesthetics widened. Stumpf's 1911 book, *Die Anfänge der Musik*, was an additional testament to the continued relevance of musical expertise as accepted musical aesthetics widened still to include non-Western music. Stumpf explicitly framed the book as a combination of contemporary work in ethnology (*Völkerkunde*), comparative musicology, and experimental psychology toward the determination of the origins of music.[70] Like Schenker, he diverged from the recent theories advanced by the likes of Charles Darwin, Herbert Spencer, and Karl Bücher, and instead located the origins of music in primitive song.[71] He also mobilized his theory of tone fusion as codetermining the development of music. By maintaining tone fusion, Stumpf's project remained a psychophysical one.

Stumpf's discussion relied heavily on the work of Hornbostel. Indeed, Hornbostel's studies of polyphonic music among primitive people (*Naturvölker*), as well as his experiments on tone distance judgment, supported Stumpf's claims locating the origins of music in song.[72] In his discussion of Hornbostel and Abraham's tone distance judgment experiments, Stumpf explained that they built on and confirmed the very assumption previously made by Fechner and Weber that the judged middle tone was geometric, the ratio of tone frequencies, and not arithmetic difference.[73] He also noted that these experiments were possible either by setting aside the "habit of our intervals" (*die Gewöhnung an unsere Intervalle*) or by experimental designs that would render these habits harmless.[74] Stumpf could not help but take one last swing at Wundt and Lorenz, continuing: "Wundt later replaced [this assumption] on the basis of the misinterpreted test results of one of his students . . . which leads to quite impossible consequences."[75]

Conclusion

The divergence of Wundt's and Stumpf's research programs following their debate in the 1890s reflected and reinforced changes that occurred both within academia and German music culture more generally. Disciplinary splits and the proliferation of new disciplines meant the frameworks within which they defined their projects were unstable. Meanwhile, the music itself had changed. Traditional music of the classical German style was no longer unchallenged as the equivalent to sound, at least not exclusive of other cultural traditions and the increasingly radical harmonies and rhythms of contemporary Western music. Wundt found little need for his observers to maintain a superior ability to recognize Western musical intervals. The older generation of natural scientists employing psychophysics to study sound sensation (Gustav Fechner, Hermann Helmholtz, and Ernst Mach, for example) were dead or engaged in other projects. New disciplines and new sounds contributed to a shifting cultural niche within which a psychophysical study of sound sensation as a study of musical aesthetics could no longer survive.

The Stumpf-Wundt debate, juxtaposed with Hanslick's sympathetic treatment of the structural listener, indicates that the status of musical expertise, both in the laboratory and on the city streets, was being renegotiated at the end of the nineteenth century. These parallel examples are suggestive of a larger trend in terms of changing conceptions of the individual experience of sound. In the music world the early writings of Hanslick, while by no means universally accepted, were widely circulated and hint at changing listening practices. The formalism outlined by Hanslick in his 1854 book provided concertgoers, already beginning to settle down and respectfully listen to music performances in relative silence, with the tools with which to reinforce *bildungsbürgerlich* behaviors. Further, by creating a hierarchy of types of listeners, topped by the aesthetic listener who experienced music through pure contemplation, Hanslick put the onus of locating the beauty of music on the individual. In this sense he also provided the concertgoers with the tools to strike out on their own and concern themselves only with their own individual, subjective experience of music. This prioritization of the individual, subjective experience of the right kind of (read: aesthetic) listener can be seen as fully realized thirty years later when Hanslick could write a semi-tongue-in-cheek article lamenting the assault of the misguided aspirations of the middling classes to acquire musical expertise on those that actually had it.

In the laboratory, studies in the psychophysics of sound sensation had been bound up with musical training and therefore musical aesthetics. Musical skill was a scientific skill. But by the end of the nineteenth century, if the Stumpf-Wundt debate is any indication, the value of musical skill had become contested. It could effect a vulnerability to observation bias and was therefore a potential menace. For Wundt, it did not reflect the listening practices of the masses. For Stumpf, who maintained a broader view of psychology than Wundt, this was precisely why it was valuable. His later work in ethnomusicology was a testament to the value of musical skill. Because Wundt's Leipzig laboratory would go on to dominate the field of experimental psychology, the role of musical expertise in the study of sound sensation quietly disappeared.

The language used in both the Hanslick and Stumpf-Wundt examples is noteworthy: music-*infected* consciousness, piano *pestilence*. Such terms laid bare the unstable status of musical skill. Musical expertise was simultaneously a means of access—to the beautiful in music, to new insights about tone differentiation—and a means of exclusion. On the city streets, it separated the pure aesthetic listener from the cacophony of talentless dilettantes. In the laboratory, it separated Stumpf from the cascade of data generated by the researchers of Wundt's Leipzig laboratory. It separated the individual from the masses.

This disappearance of the aesthetic dimension of psychophysics was reflected in and reinforced by the professionalization of such disciplines as experimental psychology and musicology as well as by the proliferation of new and different musical aesthetics at the beginning of the twentieth century. In this sense, the cultural niche changed. And yet, in the new subfield of ethnomusicology pioneered by Carl Stumpf, Erich Moritz von Hornbostel, Béla Bartók, and others, the common project of sound sensation studies and musical aesthetics framed in terms of subjective listening and historicism, lived on, the legacy of the very rich and very fruitful collaboration of music and science in the nineteenth century.

In this work I have shown how the psychophysical study of sound sensation was an aesthetic project from its very origins. And necessarily so; the psychophysicists of the nineteenth century were so deeply steeped in the music world that they thought of sound and music interchangeably. I have therefore also labored to detail the extensive intellectual and cultural engagement between the world of science and the world of music. It was, in fact, one world. Individuals like Helmholtz or Mach moved with ease in both concert halls and laboratory spaces.

I have also delineated some of the limits of this science-music exchange. Riemann's search for harmonic undertones is presented as an example of epistemologies being developed and boundaries being defended by both science and musicology. In the end, musicology did not need the natural sciences as much as the psychophysicists needed music.

This was tangled up with another thread that has woven its way through this work: the changing value and expectation of musical skills in the psychophysical laboratory. For Helmholtz, Mach, and Stumpf, the ability to read, play, and, above all, hear music was critical for their science. Stumpf's debate with Wundt, as well as the parallel increasingly shrill tone defending formalist listening in the music world, indicates that listening skills were being redefined and their value renegotiated.

Another main narrative thread of this work is the tension and eventual reconciliation of the individual, subjective experience of sound with universal laws. These laws ranged from psychophysiological theories of sound sensation to universal laws of harmony and musical aesthetics. Again and again, this reconciliation occurred through some form of historical turn. For Riemann it was a turn away from natural scientific theories of sound sensation to historical analysis to legitimize his theory of harmonic function. For Helmholtz, accepting the historical nature of musical aesthetics, most specifically his own, made it possible to locate his physiological laws of sound sensation within a progressive development of musical aesthetics. Mach came to realize that hearing itself was historical and could change over time. This realization can be understood as an important preliminary step toward what would eventually become his historicist conception of knowledge. Mach's historicist conception of knowledge and phenomenology of science remain important today, though more for philosophical, historical, anthropological, and sociological studies of science than for science itself.

Finally, we have seen the culmination of a tension that has woven throughout this book: not only was the nineteenth-century psychophysical study of sensory perception framed in terms of musical aesthetics, but it became increasingly so. As musicology and musical aesthetics shifted to prioritize and celebrate the individual listener's experience, the psychophysicists collapsed experimental subject with experimental object, accepting that subjective, individual experience was the only avenue through which to study sound sensation. Sharing a common language of musical aesthetics, psychophysicists and musicologists also shared a common interest in, first, a very general and cultural approach but then, by the beginning of the twentieth century, the aural experience of the individual.

Rather than becoming increasingly *wissenschaftlich* and narrowly preoccupied with experiment, psychophysicists were instead very much engaging the intersection of sensory perception, music culture, and subjectivity. This study has challenged older portrayals of the fin-de-siècle as a period of incommensurability and disintegration. Instead of being increasingly sequestered, which is generally the thrust of the natural sciences at the end of the nineteenth century, the psychophysical study of sound sensation left the confines of the traditional laboratory setting and converged with several of the goals of ethnomusicology as well as early twentieth-century music composition. In the ethnomusicological research that grew out of Stumpf's laboratory, the historicism of musical aesthetics was accepted and made an object of field study. For twentieth-century composers, a historicist psychophysics of sound sensation gave structure to the intention of their works—to create for their listeners highly specific, fleeting experiences of sound, fully subjective and unbound from time and space.

One of the defining features of modernity is the alienation of self. Historians of science point to the development of new technologies, new theories of perception, the unconscious, the relativistic nature of the universe as mediators in this phenomenon. Ben Steege, in his examination of the rise in laryngoscopic study for example, finds a narrative of technological intervention leading to estrangement of the voice. The laryngoscope defamiliarized individuals' own sound from themselves.[76] It was these estranging moments and their "frustrated promise of self-presence, co-presence, or co-extensivity" that have since been found to be central themes in modernist aesthetic projects.[77] Similar arguments have certainly been made about vision and the visual arts.

This was not the case for the psychophysical study of sound sensation in the nineteenth century and especially in the period this book focuses on, 1840–1910. The psychophysical study of sound sensation, achieved through both self-observation and mediating technology, could only be reconciled with musical aesthetics if and when the individual, subjective experience of sound was embraced. The individual listener was not alienated from his or her ear. Instead, not only was the psychophysical ear but the psychophysical self fully realized.

Coda

In May 1943, in the coastal mountains of Los Angeles, Thomas Mann began work on his opus *Doctor Faustus: The Life of the German Composer Adrian Leverkühn as Told by a Friend*. The friend—Serenus Zeitblom, symbolic of German Apollonian humanism—documents Leverkühn's Faustian bargain. In exchange for creative genius, Leverkühn intentionally contracts syphilis. After achieving greatness with extraordinary, atonal compositions he descends into a Dionysian darkness brought on by the untreated spirochete.

Zeitblom and Leverkühn were born in a small town in Thuringia in the 1880s to upper-middle-class families. Their youths bore all the trappings that such status would imply. Leverkühn's father would lecture the boys on animals' various methods of mimicry and disguise. With a rotating glass circle and some sand, he would employ an old cello bow and make Chladni figures to the boys' delight. As they grew older, they attended popular lectures by the visiting musicologist, Kretzschmar. These lectures sparked a passionate interest in music for Leverkühn. He would eventually combine his fascination with harmony, counterpoint, and polyphony with his subsequent theological education into a pursuit of metaphysical secrets.

Leverkühn's Mephistophelian pact and subsequent destruction are inextricably linked to the horrific moral and humanitarian collapse of his country. The year that the composer's disease first manifests in the form of mental breakdown and paralysis—1930—also marks the first of many Brownshirt electoral victories. Composer and his world both go mad. Indeed, Zeitblom ends his story a lonely man, folding his hands and stating, "May God have mercy on your poor soul, my friend, my fatherland."[1]

Like his characters, Mann came of age at the end of the nineteenth century. He was of the generation that followed the period covered in this work and for this reason is a good way to introduce a short discussion of what came of the fruitful engagement between psychophysical studies of sound sensation and music. *Doctor Faustus* is a fictional biography and it is the story of a nation. It is also a story of music. The brilliant and new twelve-tone music developed by Leverkühn was a direct reference to Schoenberg, a fellow exile in Pacific Palisades. Mann also acknowledged that he worked closely with Theodor Adorno, one of the founders of the Frankfurt School (who also lived in Pacific Palisades during World War II, down the street from Bertolt Brecht). Indeed many of Adorno's musicological ideas come through in *Doctor Faustus*, so much so that the text is considered a resource for studying Adorno.

Adorno's ideas are most explicitly articulated in the narrator's discussion of attending, with Leverkühn, lectures given by the musicologist Kretzschmar on late Beethoven, arguing that the composer's works convey a deliberate impression of being unfinished. Musicologist Rose Subotnik has carefully examined Adorno's writings on late Beethoven and argues that Schoenberg was, for Adorno, the final manifestation of the process begun by Beethoven's late style and, as such, represented the "severing of subjective freedom from objective reality," the irreconcilability of subject and object, individual freedom and social order.[2] This irreconcilability meant that the only form of protest remaining for artists (whose goal, if they sought authenticity in their work, was to criticize society) was to remove themselves from society and create art independently. It also meant that such works maintained the exclusiveness of subject and object, acknowledging their antinomies.[3]

Subotnik explains that Adorno saw both Beethoven and Schoenberg as part of the same historical movement of Western bourgeois humanism—she argues that it is this movement itself that is the true protagonist of Mann's work.[4] This humanism made social protest through art effective but also led to the eventual and irrevocable alienation of artists from their society. Artists aware of the social implications of their impotence, Subotnik argues, would have had few options, according to Adorno, besides a descent into insanity akin to Leverkühn's.[5]

Literary theorist Edward Said extended Harold Bloom's concept of belatedness to describe Adorno's *lateness*, to remain beyond what is acceptable or normal or natural, full of memory and very aware of the present, yet incapable of moving beyond this state and instead trapped in ever-deepening lateness.[6] Death, an end, in late Beethoven (and therefore also Schoenberg) does not appear definitively but rather as irony.[7] According to Subotnik, for Adorno, Schoenberg's music marked the end of a dynamic human history and the beginning of a posthistorical world.[8] He ended his discussion of Schoenberg in *Philosophy of Modern Music* with the declaration: "Modern music sees absolute oblivion as its goal. It is the surviving message of despair from the shipwrecked."[9]

Of those shipwrecked left to survive in the posthistorical world, what did they hear? Let us turn now to the current status of sensory perception research. It maintains only faint echoes of its nineteenth-century existence as a project of both psychophysics and aesthetics. Psychophysics is now celebrated as the origins of experimental psychology, at least by experimental psychologists. As such, it has been reduced to a set of classical experiments and measurement techniques presented in introductory experimental psychology courses. A sad and ironic fate considering it was the late nineteenth-century rise of experimental psychology that marked the end of psychophysics as a possible science of aesthetics. The historicist, phenomenological, and aesthetic components—products of psychophysics' origins in Fechner's strange romanticism—that were so fruitful for the individuals examined in this work are now gone.

Scientific studies that remain interested in connections between sensory perception and aesthetics are now found in neuroscience. Advances in brain imaging technology, cortical mapping, and the like give neuroscientists a renewed confidence in their ability to observe neural processes, often in real time (in many ways the early enthusiasm over fMRI technology echoes the early enthusiasm over psychophysics in the 1840s—Finally! We will be able to see the mind at work!). This has led to a series of fascinating experiments and new insights into human perceptual processes.

David Eagleman, for example—a neuroscientist at the Baylor College of Medicine in Houston—examines the active construction of perception by individual brains, how this construction differs between brains, and the implications these processes have for individuals in society. To this end, his laboratory focuses on time perception, synesthesia (the phenomenon in which an individual experiences unexpected and multiple perceptions in response to a single sensory stimulus), and neuroscientific-evidence-based policy for the punishment and rehabilitation of criminals. In a profile in the *New Yorker*, Eagleman describes his work as directly descended from the psychophysicists of the nineteenth century.[10] Framed as studies of perception and action, I suppose it is.

Eagleman's work does not, however, address anything akin to aesthetics. There is, nevertheless, a growing subfield of neuroscience termed neuroaesthetics, led by researchers Semir Zeki and Vilayanur Ramachandran at University College London and the Universities of California at Berkeley and San Diego. They too trace the origins of their work to the late eighteenth and early nineteenth centuries, specifically the search for nervous mechanisms related to aesthetic experiences. Neuroaesthetics today describes its goals as the characterization of "the neurobiological foundations and evolutionary history of the cognitive and affective processes involved in aesthetic experiences and artistic and other creative activities."[11] Neuroaestheticists seek to understand the aesthetic experience on the level of brain function. And yet Ramachandran's interest in, say, using galvanic skin response (a measurement of electrical conductance, or basically moisture) to quantify visual judgments of art sounds like something Fechner would have tried.

A quick survey of neuroaesthetic publications and conferences reveals that the field is currently limited almost entirely to the aesthetic experience of the visual arts. The perhaps best-known neuroscientific research on the brain and music is that of Oliver Sacks. His popular book, *Musicophilia: Tales of Music and the Brain*, is a collection of case studies of the strange and wondrous relationship music has with the brain. Sacks presents us with Tony Cicoria, who, after being struck by lightning, became obsessive in his desire to listen to, then master playing, then composing Chopin-like piano music. Sacks also gives us the heartbreaking story of Clive Wearing, who suffered from an amnesia so severe that his memory was reduced to mere moments, which, because his window of sustained memory was so small, also made perception completely fleeting. Wearing's musical powers, however, remained. Performing or conducting a piece of music restored his ability to connect episodic memories. Music granted him the ability to be present, at least for as long as the piece lasted; then he would again descend into the abyss of amnesia. Certainly there is something exceptional about music, giving it a different relationship to the mechanisms of memory and perception in the brain. Sacks suggests it is both the structure of the music and the way in which "each tone entirely fills our consciousness, yet simultaneously it relates to the whole," that gives Wearing perceptual continuity.[12]

It is worth noting that all of the case studies presented by Sacks are Westerners. The music referred to is Western and is squarely within the traditional Western tonal system. Sacks offers a very narrow definition of music, a nineteenth-century one, actually. All the composers are either from the nineteenth century or were reified in the nineteenth-century reparatory movement. So that movement, in its way, continues today. Sacks's work suggests that there is something special about Western harmony and melody; it exists in a different part of the brain and is capable of surviving the traumas that destroy memory, language, and personality. This faith in the exceptionalism, if not superiority, of Western musical aesthetics certainly bears resemblances to the musical tastes of many of the psychophysicists of the nineteenth century.

The discoveries of neuroscientists can have implications for aesthetics —new understandings of synesthesia or how a jolt of lightning can grant an individual virtuosic skill on the piano—and certainly this is part of what makes their work so fascinating. These lines of research are not, however, framed from the start as combined studies of aesthetics and sensory perception. They are studies of neurons that may provide insight into aesthetics. These works may be descendants of the nineteenth-century psychophysical studies of sound sensation but they are not, in fact, the legacy.

Regardless of explanation, psychophysical, neuroscientific, or otherwise, listening practices continue to form and change. Certainly what constitutes music in Western culture has broadened. As if to confirm Adorno's sentiment that current music is the work of a posthuman species in a posthistorical time, contemporary art music has been pushed to the fringes and generally understood, as Alex Ross describes it, as an "obscure pandemonium on the outskirts of culture."[13] Jazz, rock and roll, avant-garde and minimalism in the second half of the twentieth century, the pop music of today, background music in lobbies and elevators, and that purgatorial realm of being "on hold" on the phone—myriad forms of noise compete to define individuals' listening experience. All play with and prey on fractured attention. The ear in public space is battered with human-made sounds. But technology has also liberated the individual from this cacophony. Individuals can, with the aid of personal recording and replay technology, isolate themselves with the sounds of their choice, whenever and wherever they choose.

The expectations of the nineteenth-century *bildungsbürgerlich* class, to read music, to attend concerts, to play musical instruments, resonate with fewer people today. These skills are no longer considered the hallmark of refinement and culture. Presently individuals still create and receive sound but often not with their bodies or at least not with more than a tap of their finger on the screen of an electronic device. *Kids these days*.

John Cage's postwar avant-garde piece 4′33,″ in which the performers do not generate sound with their musical instruments in the expected manner, perhaps best demonstrates the changed attitudes about what constitutes musical sound and correlative new forms of listening. The listener is the music. In this sense the individual experience of sound is complete, though not because the psychophysics of sound sensation were reconciled with musical aesthetics, but because musical sound was redefined. Which leaves us with the question: Are our ears still psychophysical?

Appendix

Musical Terms and Musical Notation

Beat (interference beat) The buzzing or pulsing that results from two waves of slight phase or pitch difference sounding together.

Comma A microtonal discrepancy important for temperament and tuning. The syntonic comma is the discrepancy between a just major third and four just perfect fifths less two octaves, an interval of 81:80. A Pythagorean comma is the discrepancy between twelve perfect fifth ratios and seven octaves, an interval of 1.0136:1.

Dominant Fifth degree of a major or minor scale. For example, in the key of C, the dominant is G.

Equal temperament A tuning system in which each semitone interval between the twelve tones that make up each octave is equivalent. See *Temperament*.

Fifth (perfect) An interval of three whole tones and one semitone (of a diatonic scale) apart. In just intonation the ratio of the two pitches is 2:3.

Fourth (perfect) An interval of two whole tones and one semitone (of a diatonic scale) apart. In just intonation the ratio of the two pitches is 4:3.

Harmonic overtones (upper partials) The set of tones whose frequencies are successive integer multiples of the fundamental (root) tone. For example, for C, the harmonic overtone series is c, g, c′, e′, g′, c″ . . .

Interval The distance between two pitches, usually according to the number of steps in a diatonic scale.

Just intonation A tuning system in which all intervals are pure (they do not beat) or of small number ratios. In practice, on normal keyboards, this means prioritizing third, fourth, and fifth intervals to be pure while sacrificing others.

Key The set of notes (of a major or minor scale) generally, though not necessarily strictly, adhered to in a musical composition. The convention developed in the early seventeenth century but was abandoned by many composers in the twentieth century.

Octave An interval of five whole tones and two semitones (of a diatonic scale) apart. The ratio of the two pitches is 2:1.

Subdominant Fourth degree of a major or minor scale. For example, in the key of C, the subdominant is F.

Temperament Adjustment in tuning of musical intervals on keyboard instruments such that pairs of notes are combined rather than treated individually (B♯ and C are made the same). This eliminates pure intervals but avoids the necessity of either an unmanageable number of keys or a retuning of the keyboard for each musical composition.

Third (perfect) An interval of two whole tones (of a diatonic scale) apart. The ratio of the two pitches is 5:4. The ratio of the two pitches of a minor third is 6:5.

Tonic The main note of a key (of a major or minor scale), after which the key is named.

Musical Notation and Pitch Table A comparison of historical pitch notation of the late eighteenth and nineteenth century (used in this work) to current scientific pitch notation as well as the frequencies in cycles per second for both just intonation and equal temperament

Historical pitch notation	Modern scientific pitch notation	Frequency in just intonation (based on A4 as 440 Hz)	Frequency in equal temperament (based on A4 as 440 Hz)
C	C2	66 Hz	~ 65.4 Hz
C	C3	132 Hz	~ 130.8 Hz
G	G3	192.5 Hz	~ 1965.9 Hz
c′ (middle C on the piano)	C4	264 Hz	~ 261.6 Hz
e′	E4	330 Hz	~ 329.6 Hz
g′	G4	385 Hz	~ 391.9 Hz
c″	C5	528 Hz	~ 523.3 Hz
d″	D5	586.6 Hz	~ 587.3 Hz

Notes

Introduction

1. Ernst Mach, "On Symmetry," in Ernst Mach, *Popular Scientific Lectures*, 3rd ed., trans. Thomas McCormack (Chicago: Open Court, 1898), 98.

2. This is a reference to Lynn Nyhart's *Biology Takes Form: Animal Morphology and the German Universities, 1800–1900* (Chicago: University of Chicago Press, 1995), 4.

3. Jonathan Crary, *Suspensions of Perception: Attention, Spectacle, and Modern Culture* (Cambridge, MA: MIT Press, 1999); Michael Hagner, "Toward a History of Attention in Culture and Science," *MLN* 118, no. 3 (April 2003): 670–687; Lorraine Daston and Peter Galison, *Objectivity* (New York: Zone Books, 2007).

4. Deborah Coen, *Vienna in the Age of Uncertainty: Science, Liberalism, and Private Life* (Chicago: University of Chicago Press, 2007); Suman Seth, *Crafting the Quantum: Arnold Sommerfeld and the Practice of Theory, 1890–1926* (Cambridge, MA: MIT Press, 2010).

5. Mitchell Ash, *Gestalt Psychology in German Culture, 1870–1967: Holism and the Quest for Objectivity* (Cambridge: Cambridge University Press, 1995); Kurt Danziger, *Constructing the Subject: Historical Origins of Psychological Research* (Cambridge: Cambridge University Press, 1990); Ulla Fix and Irene Altmann, eds., *Fechner und die Folgen außerhalb der Naturwissenschaften: Interdisziplinäres Kolloquium zum 200. Geburstag Gustav Theodor Fechner* (Tübingen: Max Niemeyer Verlag, 2003); Michael Heidelberger, *Die innere Seite der Natur: Gustav Theodor Fechners wissenschaftliche-philosophische Weltauffassung* (Frankfurt am Main: Vittorio Klostermann, 1993); William Woodward and Mitchell G. Ash, eds., *The Problematic Science: Psychology in Nineteenth-Century Thought* (New York: Praeger, 1982).

6. Gary Hatfield, "Helmholtz and Classicism: The Science of Aesthetics and the Aesthetics of Science," in David Cahan, ed., *Hermann von Helmholtz and the Foundations of Nineteenth-Century Science*, 522–558 (Berkeley: University of California Press, 1993); Timothy Lenoir, "The Eye as a Mathematician: Clinical Practice, Instrumentation, and Helmholtz's Construction of an Empiricist Theory of Vision," in David Cahan, ed.,

Hermann Helmholtz and the Foundations of Nineteenth-Century Science, 109–153 (Berkeley: University of California Press, 1993); R. Steven Turner, *In the Eye's Mind: Vision and the Helmholtz-Hering Controversy* (Princeton, NJ: Princeton University Press, 1994); Robert Brain, *The Graphic Method: Inscription, Visualization, and Measurement in Nineteenth-Century Science and Culture*, doctoral dissertation, University of California at Los Angeles, 1996).

7. Peter Galison, *Image and Logic: A Material Culture of Microphysics* (Chicago: University of Chicago Press, 1997); Pamela Smith, *The Body of the Artisan: Art and Experience in the Scientific Revolution* (Chicago: University of Chicago Press, 2004). See also the "Focus" section of *Isis* 97, a collection of essays on the theme of science and visual culture by M. Norton Wise, Pamela Smith, Iwan Rhys Morus, Jennifer Tucker, and Hannah Landecker.

8. See, for example, Carl Schorske, "Explosion in the Garden: Kokoschka and Schoenberg," in Carl Schorske, *Fin-de-Siècle Vienna: Politics and Culture*, 322–366 (New York: Vintage Books, 1981); James Johnson, *Listening in Paris: A Cultural History* (Berkeley: University of California Press, 1995).

9. Emily Thompson, *The Soundscape of Modernity: Architectural Acoustics and the Culture of Listening in America, 1900–1933* (Cambridge, MA: MIT Press, 2002).

10. Elfrieda Hiebert and Erwin Hiebert, "Musical Thought and Practice: Links to Helmholtz's *Tonempfindungen*," in Lorenz Krüger, ed., *Universalgenie Helmholtz: Rückblick nach 100 Jahren* (Berlin: Akademie Verlag, 1994): 295–311; David Pantalony, *Altered Sensations: Rudolph Koenig's Acoustical Workshop in Nineteenth-Century Paris* (Heidelberg: Springer, 2009).

11. Myles Jackson, *Harmonious Triads: Physicists, Musicians, and Instrument Makers in Nineteenth-Century Germany* (Cambridge, MA: MIT Press, 2006).

Chapter 1

1. Gustav Fechner, *Elemente der Psychophysik*, vol. 2 (Leipzig: Breitkopf und Härtel, 1889), 13.

2. Carl Stumpf, "Autobiography of Carl Stumpf," in Carl Murchison, ed., *History of Psychology in Autobiography*, vol. 1, 389–441 (Worcester, MA: Clark University Press, [1924] 1930), 395.

3. Wilhelm Wundt, *Gustav Theodor Fechner: Rede zur Feier seines hundertjährigen Geburtstages* (Leipzig: Wilhelm Engelmann, 1901), 59.

4. This generation, James elaborated, included Preyer, Wundt, Paulse, and Lasswitz (William James, "Concerning Fechner," in William James, *A Pluralistic Universe: Hibbert Lectures at Manchester College* (London: Longmans, Green & Co., 1909), 148).

5. William James, introduction to Gustav Fechner's *The Little Book of Life After Death*, trans. Mary C. Wadsworth (Boston: Weiser Books, [1904] 2005), viii–ix.

6. The very notable exceptions are the recent studies of Fechner by German historians. See Michael Heidelberger's *Die innere Seite der Natur: Gustav Theodor Fechners wissenschaftlische-philosophische Weltauffassung* (Frankfurt am Main: Vittorio Klostermann, 1993), translated by Cynthia Klohr and titled *Nature from Within: Gustav Theodor Fechner and His Psychophysical Worldview* (Pittsburgh: University of Pittsburgh Press, 2004); see also Ulla Fix and Irene Altmann, eds., *Fechner und die Folgen außerhalb der Naturwissenschaften: Interdisziplinäres Kolloquium zum 200. Geburtstag Gustav Theodor Fechners* (Tübingen: Max Niemeyer Verlag, 2003).

7. Fechner and Helmholtz corresponded over a passage of *Tonempfindungen* that had confused Fechner (Fechner Nachlass 42:2, Universitätsbibliothek Leipzig).

8. Fechner diary, July 7, 1844, and November 14, 1864 (Gustav Fechner, *Tagebücher 1828 bis 1879*, vol. 1 (Leipzig: Verlag der Sächsischen Akademie der Wissenschaften, 2004), 270, 603); November 24, 1864, and January 31, 1870 (*Tagebücher 1828 bis 1879*, vol. 2, 633–637, 902–904).

9. Fechner diary, (219) October 24, 1864 (*Tagebücher 1828 bis 1879*, vol. 1, 590).

10. Eduard Hanslick, at the time studying law in Prague, apparently became acquainted with the world of contemporary music through the *Neue Zeitschrift für Musik* and even went so far as to found a Prague chapter of the Davidsbündler.

11. Eduard Hanslick, "Robert Schumann in Endenich," *Am Ende des Jahrhunderts, 1895–1899,* 2nd ed. (Berlin: Allgemeiner Verein für Deutsche Litteratur, 1899), 317–342. In *Beyond Good and Evil: Prelude to a Philosophy of the Future*, Friedrich Nietzsche explained: "Schumann, fleeing into the 'Saxon Switzerland' of his soul, half like Werther, half like Jean Paul, certainly not like Beethoven, certainly not like Byron—his Manfred music is a mistake and misunderstanding to the point of an injustice—Schumann with his taste which was basically a *small* taste (namely a dangerous propensity, double dangerous among Germans, for quiet lyricism and sottishness of feeling), constantly walking off to withdraw shyly and retire, a noble tender-heart who wallowed in all sorts of anonymous bliss and woe, a kind of girl and *noli me tangere* from the start: this Schumann was already a merely *German* even in music, not longer a European one, as Beethoven was and, to a still greater extent, Mozart. With him German music was threatened by its greatest danger: losing *the voice for the soul of Europe* and descending to mere fatherlandishness" (Friedrich Nietzsche, *Beyond Good and Evil: Prelude to a Philosophy of the Future*, trans. Walter Kaufmann (New York: Vintage Books, [1886] 1966), 181–182).

12. See Clara Schumann, *Letters of Clara Schumann and Johannes Brahms, 1853–1896*, ed. Berthold Litzmann, vol. 2 (New York: Vienna House, 1973), 2–4. Brahms had appeared at the Schumanns' door in Leipzig in 1853 with a letter of introduction

from Joseph Joachim and became a close friend, corresponding extensively with Clara until her death. Johannes Brahms in many ways carried on the stylistic mantle of Schumann and Mendelssohn. Hanslick, offering qualified praise on first hearing Brahms perform in Vienna, described Brahms's playing as more like a composer than a virtuoso and therefore irresistible in its spiritual charm. See Eduard Hanslick, "Brahms," in Henry Pleasants, ed. and trans., *Music Criticisms, 1846–1899* (New York: Penguin, [1862] 1950), 85.

13. James Garratt, *Music, Culture and Social Reform in the Age of Wagner* (Cambridge: Cambridge University Press, 2010), 84–128.

14. Garratt, *Music, Culture and Social Reform in the Age of Wagner*, 89.

15. It should be noted that while Mendelssohn's performance of Bach's *St. Matthew Passion* was not the origin of the Bach revival, it certainly contributed to it.

16. Garratt, *Music, Culture and Social Reform in the Age of Wagner*, 90.

17. Garratt, *Music, Culture and Social Reform in the Age of Wagner*, 90.

18. Hans-Jürgen Arendt, "Gustav Theodor Fechner und Hermann Härtel—eine lebenslange Freundschaft," in Ulla Fix and Irene Altmann, eds., *Fechner und die Folgen außerhalb der Naturwissenschaften: Interdisziplinäres Kolloquium zum 200. Geburtstag Gustav Theodor Fechners*, (Tübingen: Max Niemeyer Verlag, 2003), 233–244.

19. The letters from Chladni and the Weber brothers to Härtel are located in the Handschriftabteilung of the Staatsbibliothek Berlin. The bulk of the Härtel Nachlass, including his correspondence with Joseph Joachim, is in the Musikabteilung of the same library.

20. Fechner diary, (219) November 30, 1844 (*Tagebücher 1828 bis 1879*, vol. 1, 326).

21. Arthur Schopenhauer, for example, referred to Chladni's writings on temperament in *Die Welt als Wille und Vorstellung* (*The World as Will and Representation*). Arthur Schopenhauer, E. F. J. Payne, trans., *The World as Will and Representation*, vol. 1, (New York: Dover, 1969), 266.

22. Myles Jackson, *Harmonious Triads: Physicists, Musicians, and Instrument Makers in Nineteenth-Century Germany* (Cambridge, MA: MIT Press, 2006).

23. Though initially trained in law, Chladni was also an excellent pianist and moved easily through the music world. He referred to himself as an "artist" (*Künstler*), wrote for the leading musicology journal, *Allgemeine musikalische Zeitung*, and applied his work in experimental acoustics to the invention of new musical instruments. Jackson argues that Chladni's success in acoustics was due to his ability to circulate in both the music and scientific spheres of late eighteenth- and early nineteenth-century Europe (Myles Jackson, *Harmonious Triads*, 42).

24. In *Harmonious Triads*, 111–150, Jackson presents a rigorous analysis of Wilhelm Weber's study of adiabatic phenomena in reed pipes as part of a project to improve the craft of organ making.

25. E. F. F. Chladni, "Wellenlehre, auf Experimente gegründet, oder über die Wellen tropfbarer Flüssigkeiten, mit Anwendung auf die Schall- und Lichtwellen," *Cäcilia: Eine Zeitschrift für die musikalische Welt*, 4: 189–212.

26. E. H. Weber, "De Subilitate Tactu," in E. H. Weber, *De Pulsu, Resorptione, Auditu, et Tactu* (Leipzig: Prostat Apud C. F. Koehler, 1834).

27. E. H. Weber, in Helen E. Ross and David J. Murray, eds., *E. H. Weber on the Tactile Senses*, 2nd ed. (Psychology Press: East Sussex, UK, 1996), 63.

28. Weber, *E. H. Weber on the Tactile Senses*, 64–65.

29. Weber, *E. H. Weber on the Tactile Senses*, 51.

30. Thus, two points oriented along the longitudinal axis of the arm feel closer together than two points oriented along the transverse axis of the arm (Weber, *E. H. Weber on the Tactile Senses*, 110).

31. Ultimately, it was movement that allowed individuals to distinguish between objective and subjective perception, to recognize a separate object as the sensory stimulus. Weber gave the example of locating the source of a sound. Turning the head so that the right ear is oriented toward and then away from the source allowed the individual to establish the location of the source of the sound. Further, this exercise established that the sound was external to the individual's hearing organ. Essentially—though Weber did not frame it this way—it was through conscious movement that the individual established a reality external to his or her body (Weber, *E. H. Weber on the Tactile Senses*, 150).

32. Weber, *E. H. Weber on the Tactile Senses*, 92.

33. Weber, *E. H. Weber on the Tactile Senses*, 94–95.

34. Weber, *E. H. Weber on the Tactile Senses*, 101–102.

35. This work, in part, built on previous studies by the French physicist Charles Delezenne and Wilhelm Weber.

36. Weber, *E. H. Weber on the Tactile Senses*, 125.

37. Weber, *E. H. Weber on the Tactile Senses*, 212.

38. Weber, *E. H. Weber on the Tactile Senses*, 218.

39. Gustav Fechner, *Vorschule der Aesthetik*, 3rd ed. (Leipzig: Breitkopf und Härtel, [1876] 1925), v. "Ueber einige Bilder der zweiten Leipziger Kunstausstellung" can

be found in *Kleine Schriften* (Leipzig: Breitkopf und Härtel, 1875) under the name Dr. Mises.

40. Gustav Fechner, quoted in J. E. Kuntze, *Gustav Theodor Fechner (Dr. Mises): Ein deutsches Gelehrtenleben* (Leipzig: Breitkopf und Härtel, 1892), 114–115; quoted in Heidelberger's *Nature from Within*, 48.

41. G. Stanley Hall, *Founders of Modern Psychology* (New York: Appleton, 1912), 138.

42. James, "Concerning Fechner," 147.

43. Gustav Fechner, *Zend-Avesta oder über die Dinge des Himmels und des Jenseits. Vom Standpunkt der Naturbetrachtung*, 3 Theile (Leipzig: Leopold Voß), 1–14.

44. Fechner, *Zend-Avesta*, 292–293.

45. Fechner took no firm position in relation to Immanuel Kant or Johann Gottlieb Fichte, though here his thoughts appeared to lean toward Fichte's conception of consciousness. Although not necessarily related, it is worth noting that Fechner was very close to Fichte's son, Immanuel Hermann Fichte. Certainly Fechner's later criticism of the methods of Johann Friedrich Herbart's psychology specifically as well as (in his advocacy for day-view science) of those who restrict human knowledge of the world to subjectivity, conflicted with some of the basic tenets of Kantian metaphysics. See Michael Heidelberger, *Nature from Within*, 31–35, 64–65.

46. Fechner, *Zend-Avesta*, 289–293.

47. Fechner presented his most thoroughly developed conception of day-view science in *Die Tagesansicht gegenüber die Nachtansicht* (Eschborn: Klotz, [1879] 1994).

48. Fechner, *Die Tagesansicht gegenüber die Nachtansicht*, 15, 71.

49. Heidelberger argues that this two-part critique indicates that Fechner's day-view science was contradicted by Schelling/Hegel theory and Kant/Schopenhauer theory (Heidelberger, *Nature from Within*, 73–74).

50. Fechner Nachlass 42:4, Universitätsbibliothek Leipzig.

51. This is why October 22 is International Fechner Day.

52. Fechner diary, (11) September-October, 1850 (*Tagebücher 1828 bis 1879*, vol. 1, 409).

53. Fechner diary, (11) September-October, 1850 (*Tagebücher 1828 bis 1879*, vol. 1, 410–411).

54. In the conclusion to volume two of *Elemente der Psychophysik* Fechner again noted that his psychophysical work was the outgrowth of his previous studies on the relationship between the body and soul (Fechner, *Elemente der Psychophysik*, vol. 2, 553).

55. Fechner, *Elemente der Psychophysik*, vol. 1, xi.

56. Gustav Fechner, *Elements of Psychophysics*, vol. 1, trans. Helmut Adler, ed. Davis Howes and Edwin Boring (New York: Holt, Rinehart and Winston, 1966), xxvii.

57. Fechner, *Elemente der Psychophysik*, vol. 2, 10.

58. "All discussions and investigations of psychophysics relate only to the apparent phenomena of the material and mental worlds, to a world that either appears directly through introspection or through outside observation, or that can be deduced from its appearance or grasped as a phenomenological relationship, category, association, deduction, or law" (Fechner, *Elements of Psychophysics*, vol. 1, 7).

59. "In general, we call the psychic a dependent function of the physical, and vice versa, insofar as there exists between them such a constant or lawful relationship that, from the presence and changes of one, we can deduce those of the other" (Fechner, *Elements of Psychophysics*, vol. 1, 7).

60. Fechner, *Elemente der Psychophysik*, vol. 1, 12.

61. Fechner, *Elements of Psychophysics*, vol. 1, 9.

62. "My own experiments can only be the second step, following Weber's, in attempting to solve this problem, which many must follow using new modifications of method. One can say only, from what has so far been done, that in general the observations agree quite well with the law, so that one cannot doubt its approximate or exact validity within certain limits. . . . Weber's experiments were the first to prove the law in general, but his method was not suitable for making statements about its fit. My experiments have allowed us to recognize deviations but are not sufficient to eliminate them by taking into exact account the circumstances on which they depend" (Fechner, *Elements of Psychophysics*, vol. 1, 165–166).

63. Fechner, *Elemente der Psychophysik*, vol. 1, 13.

64. Fechner, *Elemente der Psychophysik*, vol. 1, 14.

65. Erich Tremmel, "Carl Emil von Schafhäutl," *Grove Music Online*, http://www.oxfordmusiconline.com.

66. Karl von Schafhäutl, *Abhandl. d. München. Akad.* 7:501, quoted in Fechner's *Elements of Psychophysics*, vol. 1, 215.

67. Schafhäutl, *Abhandl. d. München. Akad.* 7:501, quoted in Fechner's *Elements of Psychophysics*, vol. 1, 215.

68. Fechner, *Elements of Psychophysics*, vol. 1, 151.

69. Fechner, *Elements of Psychophysics*, vol. 1, 151.

70. Fechner, *Elemente der Psychophysik*, vol. 2, 182.

71. Fechner, *Elemente der Psychophysik*, vol. 1, 14.

72. The "golden section" or "golden ratio," a ratio of proportions considered the most visually pleasing, is commonly traced to Euclid. For a line where point *C* is between point *A* and point *B*, the golden ratio would occur when *C* is located where the ratio of *AC* to *CB* is equal to the ratio of *AB* to *AC*.

73. Fechner, *Vorschule der Aesthetik*, vi.

74. See also Fechner's article, "Zur experimentellen Aesthetik, th. I," *Gesellschaft der Wissenschaften zu Leipzig* 9 (1871): 553–635.

75. Fechner, *Vorschule der Aesthetik*, 1–2.

76. Fechner, *Vorschule der Aesthetik*, vol. 1, 5.

77. Fechner, *Vorschule der Aesthetik*, vol. 1, 94.

78. Fechner, *Vorschule der Aesthetik*, vol. 1, 157–177.

79. Uta Kösser, "Fechners Ästhetik im Kontext," in Ulla Fix and Irene Altmann, eds., *Fechner und die Folgen außerhalb der Naturwissenschaften: Interdisziplinäres Kolloquium zum 200. Geburtstag Gustav Theodor Fechners* (Tübingen: Max Niemeyer Verlag, 2003), 117.

Chapter 2

1. Karl von Schafhäutl, "Moll und Dur in der Natur, und in der Geschichte der neuern und neuesten Harmonielehre," *Leipziger Allgemeine Musikalische Zeitung* 13, no. 6 (1878): col. 90.

2. By autonomous music Kramer means the late eighteenth-century emancipation of music from language combined with the evolution of the "classical" genre of symphonic and chamber music with "seemingly autonomous structures" (Lawrence Kramer, *Musical Meaning: Toward a Critical History* (Berkeley: University of California Press, 2002).

3. Kramer, *Musical Meaning*, 12–13.

4. This is the reason the 440 A is sometimes termed the "Stuttgart pitch." Scheibler chose 440 Hz because he had found it to be the mean of the frequencies for the A above middle C on pianos in Vienna.

5. Within the history of Western scales, Aristoxenes, a pupil of Aristotle, first proposed the theory of equal temperament. Equal temperament was embraced by some lutenists as early as the fifteenth century. This practice was motivated by convenience. Equal semitones allowed the same fret to mark off a diatonic semitone on one string (a B♭ on an open A string, for example) and a chromatic semitone on another (an F♯ on an open F string), commonly required in Renaissance lute music.

See Mark Lindley and Ronald Turner-Smith's *Mathematical Models of Musical Scales* (Bonn: Verlag für systematische Musikwissenschaft, 1993), 44–46.

6. Alexander Ellis, "On the History of Musical Pitch," *Journal of the Society of Arts*, 28 (1880): 293–336, 400–403; 29(1881): 109–13.,

7. Owen Jorgensen, *Tuning: Containing the Perfection of Eighteenth-Century Temperament, the Lost Art of Nineteenth-Century Temperament, and the Science of Equal Temperament* (East Lansing: Michigan State University Press, 1991), 1–7. See also Ross Duffin, *How Equal Temperament Ruined Harmony (and Why You Should Care)* (New York: Norton, 2007).

8. A harmonic interval is the distance between two pitches heard simultaneously. The number of steps between the pitches in a scale traditionally indicates this. From the pitch C up to E or down to A is a third. Further up to G or down to F is a fifth. From C to the C above or below is a perfect octave, an eighth.

9. When questioned whether an individual key had absolute character or only relative character in comparison to another, Helmholtz raised the possibility that the distinctive character of keys was due, at least in part, to a particularity of the human ear. But, at least for pianos as well as bowed and wind instruments, the more likely cause of different characters of keys, according to Helmholtz, was the way a particular key was played on the instrument. Piano keys, for instance, were struck differently depending on whether they were the black or white keys. For bowed and wind instruments the different lengths of the strings or wind chamber as a particular tone is sounded contribute to the supposed character of the key (Hermann Helmholtz, *On the Sensations of Tone as a Physiological Basis for the Theory of Music*, 2nd English ed., translated, revised, and corrected from 4th German ed., with additional notes and appendix, by Alexander Ellis (New York: Dover, [1877] 1954), 310–312).

10. Hermann Helmholtz, *Die Lehre von den Tonempfindungen als physiologische Grundlage für die Theorie der Musik*, 3rd ed. (Braunschweig: Druck und Verlag von Friedrich Vieweg und Sohn), 529.

11. Table II in Ellis, "On the History of Musical Pitch."

12. The 1889 Paris Exhibition, for example, hosted many performances of native groups. Claude Debussy apparently spent many hours at the Java exhibit captivated by the gamelan music. The experience prompted him to reevaluate the biased aesthetics of the European ear: "If one listens to [gamelan music] without being prejudiced by one's European ears, one will find a percussive charm that forces one to admit that our own music is not much more than a barbarous kind of noise more fit for a travelling circus" (Claude Debussy, *Debussy on Music*, trans. Richard Langham Smith (New York: Knopf, 1977), 278).

13. In his inaugural essay in *Berliner allgemeine musikalische Zeitung*, "Ueber die Anforderungen unserer Zeit an musikalische Kritik; in besonderm Bezuge auf diese

Zeitung," Marx articulated the need to prepare the public for a new age of music. Nearly thirty years later, he described the object of his work, *Die Musik des neunzehnten Jahrhunderts und ihre Pflege: Methode der Musik*, as a thought piece on music teachers' task of preserving and perpetuating the culture of art. See Adolf Marx, "Ueber die Anforderungen unserer Zeit an musikalische Kritik," *Berliner allgemeine musikalische Zeitung* no. 1(1824):2–4, no. 2(1824):9–11, no. 3(1824):17–19, and *Die Musik des neunzehnten Jahrhunderts und ihre Pflege: Methode der Musik* (Leipzig: Breitkopf und Härtel, 1854).

14. Scott Burnham, "The Role of Sonata Form in A. B. Marx's Theory of Form," *Journal of Music Theory* 22 (1989): 247–271 (quote on 251).

15. Adolf Marx, "Form in Music," 55-90, in Scott Burnham, ed. and trans., *Musical Form in the Age of Beethoven: Selected Writings on Theory and Method* (Cambridge: Cambridge University Press, [1856] 2006), 56.

16. Marx, "Form in Music," 60–61.

17. Marx, "Form in Music," 58.

18. Marx, "Form in Music," 65.

19. Burnham, "The Role of Sonata Form in A. B. Marx's Theory of Form," 260.

20. Burnham, "The Role of Sonata Form in A. B. Marx's Theory of Form," 263.

21. A music-theoretical ontogeny recapitulating phylogeny (Burnham, "The Role of Sonata Form in A. B. Marx's Theory of Form," 264).

22. Burnham, "The Role of Sonata Form in A. B. Marx's Theory of Form," 264–265.

23. Sanna Pederson, "A. B. Marx, Berlin Concert Life, and German National Identity," *19th-Century Music* 18, no. 2 (1994): 87–107 (especially 97–98).

24. Adolf Marx, "Standpunkt der Zeitung," *Berliner allgemeine musikalische Zeitung* 3, no. 52 (1826): 421–424.

25. See the introduction to Scott Burnham, ed. and trans., *Musical Form in the Age of Beethoven: Selected Writings on Theory and Method* (Cambridge: Cambridge University Press, 2006). See also Sanna Pederson, "A. B. Marx, Berlin Concert Life, and German National Identity."

26. Adolf Marx, *The Universal School of Music: A Manual for Teachers and Students in Every Branch of Musical Art*, trans. from the 5th ed. of the original by A. H. Wehrman (London: Robert Cocks and Co., 1852), ix.

27. "The vital question for our art and its influence on the morality and the views of the people is simply this: whether its spiritual or its sensuous side is to prevail; whether it is to purify and refresh heart and soul through its inherent spiritual

power, enriching the spirit with immortal treasures, soaring aloft to thoughts and premonitions of all that is highest and eternal—or whether, void of that holy power, it is to weaken and enervate the spirit and disposition, burying them in the billows of a narcotic sensuousness and thoughtlessness that dissolves and destroys all that is upright and noble" (Adolf Marx, "The Old School of Music in Conflict with our Times," 17–34, in Scott Burnham, ed. and trans., *Musical Form in the Age of Beethoven: Selected Writings on Theory and Method* (Cambridge: Cambridge University Press, [1841] 2006), 18).

28. Marx, "The Old School of Music," 18–19.

29. Marx, "The Old School of Music," 22.

30. Sanna Pederson, "A. B. Marx, Berlin Concert Life, and German National Identity," 105.

31. "To the question: What is to be expressed with all this material? the answer will be: Musical ideas. Now, a musical idea reproduced in its entirety is not only an object of intrinsic beauty but also an end in itself, and not a means for representing feelings and thoughts. The essence of music is sound and motion" (Eduard Hanslick, *The Beautiful in Music*, ed. Morris Weitz, trans. Gustav Cohen (New York: Liberal Arts Press, [1854] 1957), 48).

32. Hanslick, *The Beautiful in Music*, 49.

33. There was also a parallel growing interest in the emotional experience of music during this period. See Hansjakob Ziemer's *Die Moderne hören: Das Konzert als urbanes Forum 1890–1940* (Frankfurt: Campus, 2008).

34. He later presented a delightful example that more fully articulated his paradoxical understanding of form as both empty structure and the infinite spirit within the structure: "All compositions are accordingly divided into full and empty 'champagne bottles'; musical 'champagne,' however, has the peculiarity of developing with the bottle" (Hanslick, *The Beautiful in Music*, 50–53).

35. "The form (the musical structure) is the real substance (subject) of music—in fact, is the music itself, in antithesis to the feeling, its alleged subject, which can be called neither its subject nor its form, but simply the effect produced" (Hanslick, *The Beautiful in Music*, 92).

36. Both Carl Dahlhaus and Thomas S. Grey have argued that much of the contemporary criticism of Hanslick was rooted in a misunderstanding of his work. See Carl Dahlhaus, *Esthetics of Music*, trans. William W. Austin (Cambridge: Cambridge University Press, [1982] 1995), 52–57. See also Thomas S. Grey, "Eduard Hanslick," *Grove Music Online*, 2012, http://www.oxfordmusiconline.com.

37. Hanslick, *The Beautiful in Music*, 51.

38. Hanslick, *The Beautiful in Music*, 51.

39. Hanslick, *The Beautiful in Music*, 51.

40. "A 'philosophic foundation of music' would first of all require us, then, to determine the definite conceptions which are invariably connected with each musical element and the nature of this connection. The double requirement of a strictly scientific framework and an extremely comprehensive casuistry renders it a most arduous though not an impossible task, unless, indeed, our ideal is that of a science of music in the sense in which chemistry and physiology are sciences!" (Hanslick, *The Beautiful in Music*, 57).

41. Hanslick, *The Beautiful in Music*, 83.

42. "Now, on comparing *music* with those arts, it is obvious that Nature has provided no model capable of becoming its subject matter. *There is nothing beautiful in Nature as far as music is concerned*" (Hanslick, *The Beautiful in Music*, 112).

43. Hanslick, *The Beautiful in Music*, 110.

44. "That which is regarded as purely material, as the transmitting medium, is the product of a thinking mind, whereas that which is presumed to be the subject —the emotional effect—belongs to the physical properties of sound, the greater part of which is governed by physiological laws" (Hanslick, *The Beautiful in Music*, 92).

45. "There is no art which, like music, uses up so quickly such a variety of forms. Modulations, cadences, intervals, and harmonious progressions become so hackneyed within fifty, nay, thirty years, that a truly original composer cannot well employ them any longer, and is thus compelled to think of a new musical phraseology. Of a great number of compositions which rose far above the trivialities of their day, it would be quite correct to say that there *was* a time when they were beautiful" (Hanslick, *The Beautiful in Music*, 58).

46. To some extent, this explains why Hanslick refused to study or analyze any works prior to the eighteenth century.

47. Dahlhaus, *Esthetics of Music*, 54.

48. "The number of those who thus listen to, or rather feel, music is very considerable. While in a state of passive receptivity, they suffer only what is elemental in music to affect them, and thus pass into a vague 'supersensible' excitement of the senses produced by the general drift of the composition. Their attitude toward music is not an observant but a pathological one. They are, as it were, in a state of waking dreaminess and lost in a sounding nullity, their mind is constantly on the rack of suspense and expectancy" (Hanslick, *The Beautiful in Music*, 89–90).

49. Hanslick, *The Beautiful in Music*, 91.

50. Hanslick, *The Beautiful in Music*, 91.

51. Hanslick, *The Beautiful in Music*, 92.

52. Hanslick, *The Beautiful in Music*, 99.

53. Rose Subotnik, *Developing Variations: Style and Ideology in Western Music* (Minneapolis: University of Minnesota Press, 1991), 279.

54. Subotnik, *Developing Variations*, 277–278.

55. Subotnik, *Developing Variations*, 280–283.

56. By the end of his career, Riemann had published over fifty works. His writings can be divided loosely into three categories: pedagogical, theoretical, and historical. The first category consisted of practical texts on harmony and modulation, composition, orchestration, and the like, oriented toward the needs of musicians and students of composition. Riemann's theoretical works were articulations of his speculative harmonic system, an attempt to explain the meaning of harmonies, the relation of these harmonies to each other, and the evolution of what he believed were their natural laws. In his historical treatises, written in the latter part of his career, Riemann traced the history of music theory and through analyses of various pieces attempted to demonstrate universal laws of musical development. Michael Arnzt argues that the prolific writings were in fact motivated by Riemann's lack of regular income and need for money to support his family. The irony is that the volume of Riemann's writings overwhelmed his contemporaries' writings and significantly contributed to his legacy as the "father of musicology" (Michael Arnzt, "'Nehmen Sie Riemann ernst?' Zur Bedeutung Hugo Riemanns für die Emanzipation der Musik," in Tatjana Böhme-Mehner and Klaus Mehner, eds., *Hugo Riemann (1849–1919): Musikwissenschaftler mit Universalanspruch* (Vienna: Bölhau, 2001), 9–16).

57. For example, Riemann discussed the works of Fechner, Helmholtz, Hanslick, and Lotze in great deal in "Der gegenwärtige Stand der musikalischen Aesthetik," in Hugo Riemann, *Präludien und Studien: Gesammelte Aufsätze zur Aesthetik, Theorie, und Geschichte der Musik*, vol. 2 (Leipzig: Hermann Seemann, Nachfolger, [1878] 1900).

58. Hugo Riemann, "Der gegenwärtige Stand der musikalischen Aesthetik," 54.

59. Trevor Pearce points out that the changing title of Riemann's dissertation makes significantly more sense when considered in light of the fact that Riemann trained in logic with Christoph Sigwart, which contributed to Riemann's understanding of musical hearing as a logical activity (Trevor Pearce, "Tonal Functions and Active Synthesis: Hugo Riemann, German Psychology, and Kantian Epistemology," *Intégral* 22 (2008): 81–116 (especially 94)).

60. Jean-Phillipe Rameau, one of the first to attempt to unify and rationalize music according to acoustical principles, established that the lowest pitch of the triad and 7th chord was the "fundamental sound." The series of fundamental sounds in a succession of chords would reveal the "fundamental bass." In this way the fundamental bass could be an effective means of analyzing localized harmonic connections. Rameau later

presented a theory of sound as the composite of overtones. In his 1737 work, *Génération harmonique*, he included a discussion of an undertone to explain the generation of minor chords. For extended discussions of how Riemann read and misread Rameau, see Carl Dahlhaus, "War Zarlino Dualist?", *Die Musikforschung* 10 (1957): 286–290; Scott Burnham, "Method and Motivation in Hugo Riemann's History of Music Theory," *Music Theory Spectrum* 14 (1992): 4–9; Alexander Rehding, *Hugo Riemann and the Birth of Modern Musical Thought* (Cambridge: Cambridge University Press, 2003), 93–98.

61. Pearce, "Tonal Functions and Active Synthesis," 86.

62. Sadly, this title was lost in the translation of the English edition of the work, titled *Catechism of Musical Aesthetics* (London: Augener and Co., 1895).

63. Hugo Riemann, *Wie hören wir Musik?: Grundlinien der Musikästhetik* (Leipzig: Max Hesse, 1888), 3.

64. Riemann, *Wie hören wir Musik?*, 32.

65. Riemann, *Wie hören wir Musik?*, 44.

66. Riemann, *Wie hören wir Musik?*, 44.

67. Pearce, "Tonal Functions and Active Synthesis."

68. Arthur von Oettingen, *Harmoniesystem in dualer Entwickelung: Studien zur Theorie der Musik* (Dorpat: W. Gläser, 1866).

69. In 1894, following the russification of Dorpat, Oettingen moved to Leipzig to a position of honorary professor at the university, where he remained until 1919.

70. Hain Tankler, "Tartu and the Wider World: Through International Contacts to the Peaks of Science," *Revue de la Maison Française d'Oxford* 1, no. 2 (2003): 69–94. Tankler references Oettingen's article, "Der Wind-Componenten-Integrator," *Repertorium für Meteorologie* 5, no. 10 (1877): 1–51.

71. Oettingen did distinguish the *tonality* (tonic) of major triads from the *phonality* (phonic) of minor triads as the property of intervals and chords having the same fundamental tone versus the property of intervals and chords having a common overtone. But this distinction between types of harmonics was not as critical to establishing the dualistic nature of his harmonic system as it was for Riemann.

72. A triad is a three-note chord made up of two superimposed third intervals. In a major triad (C-E-G, for example), the lower third is major and the upper third is minor. In a minor triad (C-E^{b}-G, for example), the lower third is minor and the upper third is major.

73. Both Riemann and Oettingen drew on Moritz Hauptmann's influential treatise, *Die Natur der Harmonik und Metrik* (Leipzig: Breitkopf und Härtel, 1853), which proposed that the dualism of harmony and meter grew from one single, natural principle.

74. Hugo Riemann, *Geschichte der Musiktheorie im IX.–XIX. Jahrhundert,* 3rd ed. (Berlin: Georg Olms Verlagsbuchhandlung Hildesheim, [1898] 1961), 517–518.

75. Riemann argued that the first shared upper partial of a triad chord—C-E-G, for example—would stand out prominently to a listener. So, using the example of the C-E-G major triad, the sounding C, E, and G would each have their own set of upper partials. Of these three series of upper partials, the first they would have in common would be the G two octaves higher (the g′). Helmholtz would claim that this common partial occurring so (relatively) early in the sounding notes' respective upper partial series, their shared sound waves constructively interfering, was the reason this chord was perceived as consonant by a listener. Riemann elaborated on this by claiming that this common partial would stand out to a listener, not an unreasonable claim given that the wave interference would have increased the amplitude. Riemann understood this phenomenon to be further support for his belief that the third was the building block of harmonic dualism (Hugo Riemann, "Die Natur der Harmonik," *Waldersees Sammlung musikalischer Vorträge* 4 (1882): 159–190).

76. Hugo Riemann, *Musikalische Logik: Hauptzüge der physiologischen und psychologischen Begründung unseres Musiksystems* (Leipzig: C. F. Kahnt, 1874). Rehding argues that Riemann misread Oettingen. See also Richard Münnich, "Von Entwicklung der Riemannschen Harmonielehre und ihrem Verhältnis zu Oettingen und Stumpf," in Carl Mennicke, ed., *Riemann-Festschrift,* 60–76 (Leipzig: Max Hesse, 1909).

77. Riemann did, however, also highlight the critical role of the instruments and musicians themselves in creating consonant sound.

78. It might be asserted that the mirroring of undertones and overtones in Riemann's theory of harmonic dualism reflected his predilection for symmetry—that harmonic dualism was a means of reifying symmetry as a fundamental, universal quality of music. Alexander Rehding presents a convincing argument refuting this. He examines an analysis done by Riemann of Beethoven's *Waldstein* Sonata and shows that Riemann ignored many opportunities to emphasize the symmetry of the work and instead highlighted other features, most notably its ordinariness. Riemann was, in fact, much more enthusiastic about the dominant role of the third (interval) in harmonic dualism than symmetry (Rehding, *Hugo Riemann and the Birth of Modern Musical Thought*).

79. Riemann, *Musikalische Logik,* 29. Riemann developed numerous iterations of this "grid of harmonic relations." Another nice example can be seen in his *Grosse Kompositionslehre,* vol. 1 (Stuttgart: W. Spemann, 1902.

80. The term *undertone* was commonly understood (by Oettingen as well) to mean the fundamental tone of a series of overtones, *not* a separate series of partial tones as defined by Riemann.

81. A popular song by Friedrich Wilhelm Kücken (op. 17, no. 1).

82. Riemann, *Musikalische Logik*, 6.

83. "The ear makes a sound, a tone with all overtones, none of which is amplified —vibrating not only the overtones and undertones of the main tone, but also the overtones of overtones with corresponding fibers" (Riemann, *Musikalische Logik*, 14).

84. Riemann *Musikalische Logik*, 14.

85. Alexander Ellis, "Appendix XI: Vibration of the Membrana Basilaris in the Cochlea," to Hermann Helmholtz's *On the Sensations of Tone as a Physiological Basis for the Theory of* Music, 2nd English ed., translated, revised and corrected from 4th German ed., with additional notes and appendix, by Alexander Ellis (New York: Dover, [1877] 1954), 411.

86. Hugo Riemann, *Musikalische Syntaxis: Grundriss einer harmonischen Satzbildungslehre* (Breitkopf und Härtel, 1877), 6–7.

87. Riemann, *Musikalische Syntaxis*, 7.

88. Riemann, *Musikalische Syntaxis*, 121.

89. Riemann, *Musikalische Syntaxis*, 121.

90. Riemann, *Musikalische Syntaxis*, 121.

91. "It is even dared to build a theory on appearance of tones that no one has heard or can hear, although our latest theorists believe they really have heard it. But we know all too well how easy it is for one to believe he hears musical tones which lie on the border of musical audibility, if one wants to hear it" (Schafhäutl, "Moll und Dur in der Natur, und in der Geschichte der neuern und neuesten Harmonielehre").

92. Schafhäutl, "Moll und Dur in der Natur, und in der Geschichte der neuern und neuesten Harmonielehre."

93. Schafhäutl, "Moll und Dur in der Natur, und in der Geschichte der neuern und neuesten Harmonielehre."

94. Schafhäutl, "Moll und Dur in der Natur, und in der Geschichte der neuern und neuesten Harmonielehre."

95. Alexander Ellis, footnote in Hermann Helmholtz's *On the Sensations of Tone as a Physiological Basis for the Theory of Music*, 365.

96. David Pantalony, *Altered Sensations: Rudolph Koenig's Acoustical Workshop in Nineteenth-Century Paris* (Heidelberg: Springer, 2009), 158–160.

97. "I have not been able to convince myself of the correctness of the fact adduced. . . . that the undertones of a tone strongly struck on the piano sound when the corresponding dampers are raised. Perhaps the author has been deceived by the circumstance that with very resonant instruments (especially the older ones) any

strong shake, and therefore probably a violent blow on the digitals will cause some one or several of the deeper strings to sound its note." The translator added that perhaps a sympathetically vibrating lower string deceived Riemann. Hermann Helmholtz, *On the Sensations of Tone as a Physiological Basis for the Theory of Music*, 4th English ed., translated, revised, and corrected from 4th German ed., with additional notes and appendix, by Alexander Ellis (London: Longmans, Green, and Co., 1912), 365 ff. See also Carl Stecker's discussion of Helmholtz's dismissal of Riemann's undertones (Carl Stecker, "Kritische Beiträge zu einigen Streitfragen in der Musikwissenschaft," *Vierteljahrsschrift für Musikwissenschaft*, 437–465).

98. Rehding highlights the efforts of Georg Capellen and Bernhard Ziehn to bait Riemann (Rehding, *Hugo Riemann and the Birth of Modern Musical Thought*, 18). Riemann did, however, acknowledge the work of Capellen in his later writings—for example, "Das Problem des harmonischen Dualismus," *Neue Zeitschrift für Musik*, no. 1 (1905): 3.

99. Hugo Riemann, "The Nature of Harmony," quoted in William Mickelsen, ed. and trans., *Hugo Riemann's Theory of Harmony and History of Music Book III* (Lincoln: University of Nebraska Press, [1882] 1977), 46.

100. Riemann, "The Nature of Harmony," 46.

101. Hugo Riemann, "Einige seltsame Noten bei Brahms und anderen," in Hugo Riemann, *Präludien und Studien: Gesammelte Aufsätze zur Aesthetik, Theorie und Geschichte der Musik*, vol. 3 (Leipzig: Hermann Seemann, [1889] 1901), 109–123. The article was originally published in *Musikalisches Wochenblatt.*

102. Riemann, "Einige seltsame Noten bei Brahms und anderen," 112.

103. Hugo Riemann, "Untertöne," in Hugo Riemann, *Hugo Riemanns Musiklexikon*, 11th ed. (Leipzig: Max Hesse, 1929), 1902.

104. Hugo Riemann, *Katechismus der Akustik (Musikwissenschaft)* (Leipzig: Hesse, 1891), 79–80.

105. Riemann, *Katechismus der Akustik (Musikwissenschaft)*.

106. Carl Stumpf, *Die Anfänge der Musik* (Leipzig: J. A. Barth, 1911), 27.

107. Rehding includes a nice discussion of the extent to which Riemann manipulated Stumpf's concept of tone fusion and later clashed with Stumpf over the term (Rehding, *Hugo Riemann and the Birth of Modern Musical Thought*, 52–54, 108–109).

108. Hugo Riemann, "Der gegenwärtige Stand der musikalischen Aesthetik," 48.

109. Indeed a monistic interpretation of Riemann's theory of harmonic function—though divorced from harmonic dualism—remained institutionalized in the halls of Central and Eastern European musicology and continues, worldwide, to be accepted today. A growing number of musicologists are, however, criticizing this acceptance, not for the flaws in Riemann's work, but because the current celebration of Riemann

ignores the undertone experiments at the core of his theories. See Scott Burnham, "Musical and Intellectual Values: Interpreting the History of Tonal Theory," *Current Musicology* 53 (1993): 76–88, as well as Rehding, *Hugo Riemann and the Birth of Modern Musical Thought.*

110. Large-scale efforts to publish complete editions of composers beginning in the middle of the nineteenth century contributed to the development of musicology. In 1851, the Bach Gesellschaft unveiled their plans for a Bach edition. This project was followed by an edition devoted to Handel in 1858, Mozart in 1876, Schubert in 1883, and Beethoven in 1884.

111. For example, excerpts from Ernst Mach's 1886 work, *Die Analyse der Empfindungen und das Verhältnis des Physischen zum Psychischen,* were reprinted in the *Vierteljahrsschrift für Musikwissenschaft.*

112. Alexander Rehding, "The Quest for the Origins of Music in Germany circa 1900," *Journal of the American Musicological Society* 53 (2000): 345–385.

113. Alfred Kelly, *Descent of Darwin: The Popularization of Darwinism in Germany, 1860–1914* (Chapel Hill: University of North Carolina Press, 1981).

114. Rehding, "The Quest for the Origins of Music in Germany circa 1900," 346–349.

115. Hugo Riemann, ed., *Sechs originale chinesische und japanische Melodien* (Leipzig: Breitkopf und Härtel, 1902).

116. Hugo Riemann, "Musical Logic," in William Mickelsen, ed. and trans., *Hugo Riemann's Theory of Harmony and History of Music Book III* (Lincoln: University of Nebraska Press, [1874] 1977), 185.

117. Riemann, *Geschichte der Musiktheorie im IX.–XIX. Jahrhundert,* 529.

118. Rehding, *Hugo Riemann and the Birth of Modern Musical Thought,* 61. Stumpf, as will be discussed in the final chapter, once similarly claimed that Wilhelm Wundt's experimental protocols were so arbitrary that they were equivalent to a vote of the masses (the presumption being that the vote of the masses was arbitrary). And, actually, the Viennese music theorist Heinrich Schenker once accused Riemann of being a democrat (Nicholas Cook, *The Schenker Project: Culture, Race and Music Theory in Fin-de-Siècle Vienna* (Oxford: Oxford University Press, 2007)). The implied negative connotations of such zingers suggest a deep-seated anxiety about liberal democracy at the end of the nineteenth century.

119. Riemann, "Musical Logic," 185.

120. Hugo Riemann, *Grundriss der Musikwissenschaft* (Leipzig: Quelle & Meyer, 1908), 8–9.

121. Alexander Rehding makes the convincing argument that, despite the many disparate theories and methodological approaches, the proto-musicologists of late

nineteenth-century Germany united in a quest for the origins of Western tonality (Rehding, "The Quest for the Origins of Music in Germany circa 1900").

122. Rehding, *Hugo Riemann and the Birth of Modern Musical Thought*, 183.

Chapter 3

1. Hermann Helmholtz, letter dated November 5, 1838 (no. 6), in Herman Helmholtz, *Letters of Hermann von Helmholtz to His Parents: The Medical Education of a German Scientist, 1837–1846*, ed. David Cahan (Stuttgart: Franz Steiner Verlag, 1993), 51.

2. George Bernard Shaw, *Music in London, 1890–1894*, vol. 2 (London: Constable, 1932), 276–277.

3. M. Norton Wise, "A Spectacle for the Gods," in *Bourgeois Berlin and Laboratory Science*, unpublished manuscript, University of California at Los Angeles, 2012.

4. The 1791 introduction of the single-piece cast iron frame marked the beginning of a century of innovation. The flurry of design innovation was fueled in part by a growing and robust piano market that mirrored economic growth and stability, especially for the middle classes.

5. Hermann Helmholtz, letter dated November 5, 1838 (no. 6), in Helmholtz, *Letters of Hermann von Helmholtz to His Parents*, 51.

6. Some very good historical work has explored the development and reception of Helmholtz's *On the Sensations of Tone*; Elfrieda Hiebert and Erwin Hiebert present particularly useful analyses. Other scholars have traced how the link between Helmholtz's effort to bridge the natural sciences to the moral sciences was fueled by his belief in an inductive foundation rooted in empiricism. But all of these studies begin with Helmholtz's physics and physiology and then proceed to examine how his ideas related to and were received by contemporary musicians and music theorists. Current scholars have reduced his work with instruments to purely acoustic exercises and treated the piano largely as a machine. See Gary Hatfield, "Helmholtz and Classicism: The Science of Aesthetics and the Aesthetics of Science," in David Cahan, ed., *Hermann von Helmholtz and the Foundations of Nineteenth-Century Science*, 522–558 (Berkeley: University of California Press, 1993); Elfrieda Hiebert and Erwin Hiebert, "Musical Thought and Practice: Links to Helmholtz's *Tonempfindungen*," in Lorenz Krüger, ed., *Universalgenie Helmholtz: Rückblick nach 100 Jahren*, 295–314 (Berlin: Akademie Verlag, 1994); Elfrieda Hiebert, "Helmholtz's Musical Acoustics: Incentive for Practical Techniques in Pedaling and Touch at the Piano," *Papers Read at the IMS Intercongressional Symposium*, 425–433 (Budapest: Liszt Ferenc Academy of Music, 2004); Edward Jurkowitz, "Helmholtz and the Liberal Unification of Science," *Historical Studies in the Physical Science* 32 (2003): 291–317.

7. M. Norton Wise, "Making Visible," *Isis* 97 (2006): 75–82.

8. Most recent scholarship has, however, focused on the vision sense only. Norton Wise argues that making things visible makes them real and therefore encourages us to recognize that visualization in science is not merely illustration but can be argument itself. Pamela Smith presents the idea of fifteenth- and sixteenth-century artisanal knowledge in which image making is knowledge producing and argues further, in cases where images were imitations of the processes of nature, that visual art and science can be regarded as sharing the same goals. See Wise, "Making Visible"; Peter Galison, *Image and Logic: A Material Culture of Microphysics* (Chicago: University of Chicago Press, 1997); Pamela Smith, "Art, Science, and Visual Culture in Early Modern Europe," *Isis* 97 (2006): 83–100. See also Pamela Smith's book, *The Body of the Artisan: Art and Experience in the Scientific Revolution* (Chicago: University of Chicago Press, 2004).

9. Historians Frederic Holmes and Kathryn Olesko argue that the curves produced by instruments in Helmholtz's studies of muscle contraction in frogs served descriptive, demonstrative purposes that permitted a more comprehensible translation of Nature (Frederic Holmes and Kathryn Olesko, "The Images of Precision: Helmholtz and the Graphical Method in Physiology," in M. Norton Wise, ed., *The Values of Precision*, 198–221 (Princeton, NJ: Princeton University Press, 1995)). Gary Hatfield similarly finds idealization and ideal types centrally located in Helmholtz's efforts to bring his studies of visual and acoustic sensation to bear on aesthetics, suggesting that Helmholtz maintained a larger classicist goal of revealing universal truths. Hatfield also locates romantic themes in Helmholtz's work. Helmholtz acknowledged that imagination played the dominant role in artistic and aesthetic cognition. But Hatfield maintains that these romantic themes were secondary to the classicist ones. Imagination, for Helmholtz, was merely a filter for the lawlike regularity of an ideal type (Hatfield, "Helmholtz and Classicism," 524). Norton Wise, in contrast to Holmes and Olesko, has further argued that Helmholtz's aesthetic classicism can be seen in his conception of *Form* in his work on the temporal processes of muscle contraction (Wise, "A Spectacle for the Gods").

10. Gary Hatfied, *The Natural and the Normative: Theories of Spatial Perception from Kant to Helmholtz* (Cambridge, MA: MIT Press, 1990); Michael Heidelberger, "Force, Law, and Experiment: The Evolution of Helmholtz's Philosophy of Science," in David Cahan, ed., *Hermann von Helmholtz and the Foundations of Nineteenth-Century Science* 522–558 (Berkeley: University of California Press, 1993); R. Steven Turner, *In the Eye's Mind: Vision and the Helmholtz-Hering Controversy* (Princeton, NJ: Princeton University Press, 1994); R. Steven Turner, "Consensus and Controversy: Helmholtz on the Visual Perception of Space," in David Cahan, ed., *Hermann von Helmholtz and the Foundations of Nineteenth-Century Science* 154-204 (Berkeley: University of California Press, 1993).

11. Lenoir describes how Helmholtz understood the vision process to be an experiment constantly performed by the brain, in which space is a tool acquired through

a laborious learning exercise. Lenoir argues that this understanding of vision was part of a larger movement to develop realism in both physiological optics and painting in mid- to late nineteenth-century Germany. He argues that the individuals engaged in these projects shared a "common cultural horizon" and, in the wake of the failures of the revolutionary 1840s, sought to stabilize the factual basis of the world through a new concept of realism as a form of life constituted through practice (Timothy Lenoir, "The Politics of Vision: Optics, Painting, and Ideology in Germany, 1845–1895," in Timothy Lenoir, *Instituting Science: The Cultural Production of Scientific Disciplines* 131–178 (Stanford, CA: Stanford University Press, 1997)).

12. For example, Helmholtz mentioned neither psychophysics nor Fechner in *Tonempfindungen.* He was, however, aware of Fechner's work. The two men even had a brief correspondence in which Helmholtz clarified some points of his sound sensation theories for Fechner. A short letter from Helmholtz to Fechner, dated March 12, 1872, is included in the Fechner Nachlass at the Universitätbibliothek Leipzig.

13. Helmholtz's father, Ferdinand, at the time the subrector of the Potsdam Gymnasium, had published an article in 1837 in which he explained that it was through careful, practiced drawing that the power of the idea was awakened and expressed, that the Gymnasium curriculum should be structured around this gradual acquisition and understanding of the beautiful (Ferdinand Helmholtz, "Die Wichtigkeit der allgemeine Erziehung für das Schöne," *Zu der öffentlichen Prüfung der Zöglinge zdes hiesigen Königlichen Gymnasiums den 21sten und 22sten März laden ganz ergebenst ein Director und Lehrercollegium,* 1–44 (Potsdam: Decker'schen Geheimen Oberhofbuchdruckerei-Establissement, 1837)).

14. Ferdinand Helmholtz, letter dated November 2, 1838 (no. 2), in Helmholtz, *Letters of Hermann von Helmholtz to His Parents*; English translation from Leo Koenigsberger, *Hermann von Helmholtz,* trans. Frances A. Welby (Oxford: Clarendon Press, 1906), 14. This belief in music as a national cultural resource, as central to the moral development and discipline of the children of the *Bildungsbürgertum,* was particularly prevalent throughout the first half of the nineteenth century. See Myles Jackson, *Harmonious Triads: Physicists, Musicians, and Instrument Makers in Nineteenth-Century Germany* (Cambridge, MA: MIT Press, 2006), 245–248.

15. Hermann Helmholtz, letter dated November 5, 1838 (no. 6), in Helmholtz, *Letters of Hermann von Helmholtz to His Parents,* 51.

16. Hermann Helmholtz, letter dated December 1, 1838 (no. 7), in Helmholtz, *Letters of Hermann von Helmholtz to His Parents,* 53.

17. "For the first two days I had to content myself with Uncle's sheet music, which aside from a Mozart sonata, consisted of nothing but admirable works by Strauss, Lanner, Czerny, Hünten, Auber, Ross and Bellini, etc. etc. etc.; I pursued them all consecutively, but it was in the end completely awful due to the fact that I always return to the Mozart sonatas and Cramer études in order to strengthen my intellectual

stomach. On Wednesday I played a second Mozart sonata at Schmidt's, and yesterday I met Wiggers playing there, I told him my troubles, and he said I should come with him and he would give me sheet music. Since I did so I nearly collapsed (poetic hyperbole) under the heap but most are piano concertos and other artificialities of Weber, Moscheles, Kramer, etc. Only one book of Mozart's and Beethoven's Sonata are among them" (Hermann Helmholtz, letter dated August 23, 1839 (no. 13), in Helmholtz, *Letters of Hermann von Helmholtz to His Parents*, 68).

18. Helmholtz eventually came to believe that the modification of tones on the piano was dictated entirely by the mechanics of the instrument and that performers could only control the speed with which the hammer struck the string, nothing more. See Hiebert, "Helmholtz's Musical Acoustics."

19. Hermann Helmholtz, "On the Motion of the Strings of a Violin," Communicated by William Thomson, *Proceedings of the Philosophical Society of Glasgow* 17–21 (Glasgow: Philosophical Society, 1860).

20. Hermann Helmholtz, *On the Sensations of Tone as a Physiological Basis for the Theory of Music*, 2nd English ed., translated, revised and corrected from 4th German ed., with additional notes and appendix, by Alexander Ellis (New York: Dover, [1877] 1954), 74.

21. Hermann Helmholtz, "The Aim and Progress of Physical Science: An Opening Address Delivered at the Naturforscher Versammlung in Innsbrück," in Herman Helmholtz, *Popular Lectures on Scientific Subjects*, trans. E. Atkinson (Longmans, Green, and Co., [1869] 1904), 321.

22. In Koenigsberger, *Hermann von Helmholtz*, 232–233. Helmholtz is referring to Beethoven's ballet *Prometheus*. The Mendelssohn Overture is likely the Opus 21 overture to his *Midsummer's Night Dream*. The concert also included a Bach chorus as well as Handel's *Hallelujah Chorus*.

23. "But this increased theoretical accuracy would practically be illusory, because now the whole error of 1/885 lies with the fifth to the very limits of what a trained ear can distinguish" (Hermann Helmholtz, "Über musikalische Temperatur," Paper presented on November 23, 1860, *Naturh.-med. Verein in Heidelberg* 2 (1859–1862): 73–75), quotation on 75.

24. Scott Burnham, "The Role of Sonata Form in A. B. Marx's Theory of Form," *Journal of Music Theory* 22, no. 2 (1989): 264–265.

25. Eduard Hanslick, *Vom Musikalisch-Schönen: Ein Beitrag zur Revision der Ästhetik der Tonkunst*, 12th ed. (Leipzig: Breitkopf und Härtel, [1854] 1918), 59.

26. Hanslick, *Vom Musikalisch-Schönen*, 62.

27. Carl Dahlhaus, *Nineteenth-Century Music*, trans. J. Bradford Robinson (Berkeley: University of California Press, 1989), 50.

28. William Weber, *Music and the Middle Class: The Social Structure of Concert Life in London, Paris, and Vienna* (New York: Holmes and Meier, 1975). Pushing Weber's concept further, Dahlhaus argues that the dichotomy between Italo-French opera and German classical instrumental music (reified in the contrasting of Rossini and Beethoven) was at the very roots of the nineteenth-century concept of music (Dahlhaus, *Nineteenth-Century Music*, 9–11).

29. Weber, *Music and the Middle Class*, 19–20.

30. Hanslick, *Vom Musikalisch-Schönen*, 62. Also see Jackson, *Harmonious Triads*, 232–233; Dahlhaus, *Nineteenth-Century Music*, 50.

31. Celia Applegate and Pamela Potter, "Germans as the 'People of Music': Genealogy of an Identity," in Celia Applegate and Pamela Potter, eds., *Music and German National Identity*, 1–35 (Chicago: University of Chicago Press, 2002). See also Celia Applegate's *Bach in Berlin: Nation and Culture in Mendelssohn's Revival of the St. Matthew Passion* (Ithaca, NY: Cornell University Press, 2005), in which she argues that the 1829 performance of *St. Matthew Passion* in Berlin was a moment of consolidation, transformation, and self-realization in collective life that suggested such moments were perhaps only possible through musical performance.

32. Sanna Pederson, "A. B. Marx, Berlin Concert Life, and German National Identity," *19th-Century Music* 17 (1994–1995): 87–107 (especially 88).

33. Wise, "A Spectacle for the Gods."

34. Wise, "A Spectacle for the Gods."

35. Wise, "A Spectacle for the Gods."

36. In "On the Physiological Causes of Harmony in Music," in Helmholtz's *Popular Lectures on Scientific Subjects* (London: Longmans, Green, and Co., [1857] 1903), 53–54, he described music as "immaterial, evanescent, and tender creator of incalculable and indescribable states of consciousness." Believing that painting or sculpture borrowed material from human experience and usually attempted to describe or imitate nature, Helmholtz thought that they could also be investigated through scientific art criticism with respect to their correctness and truth. Helmholtz's primacy of music over all other forms of art was not uncommon and was in fact the norm for the time period.

37. "Just as in the rolling ocean, this movement, rhythmically repeated, and yet ever varying, rivets our attention and hurries us along. But whereas in the sea, blind physical forces alone are at work, and hence the final impression on the spectator's mind is nothing but solitude—in a musical work of art the movement follows the outflow of the artist's own emotions. Now gently gliding, now gracefully leaping, now violently stirred, penetrated or laboriously contending with the natural expression of passion, the stream of sound, in primitive vivacity, bears over into the hearer's

soul unimagined moods which the artist has overheard from his own, and finally raises him up to that repose of everlasting beauty, of which God has allowed but few of his elect favorites to be the heralds" (Helmholtz, "On the Physiological Causes of Harmony in Music," 93).

38. "Many many greetings out to Dahlem to my Fräulein, for whom I have so much love, so much love, as the sky is high and the sea is deep, and whom I can not stop loving, even if the sky and sea should be long since perished, to my dear Fräulein Nightingale, with the sweet and low songs of *Alceste* and dainty white hands, with whom it is so nice to stroll in large and beautiful Beethovenian (*Beethovenschen*) and other groves and gardens. So many good wishes from happy men to whom heaven has given such a little bride to protect and to love through the whole of life" (Hermann Helmholtz, letter dated June 20, 1848 (no. 16), in Richard L. Kremer, ed., *Letters of Hermann von Helmholtz to His Wife, 1847–1859* (Stuttgart: Franz Steiner Verlag, 1990), 41). *Alceste,* Chrisoph Martin Wieland's text, was set to opera by Anton Schweitzer in 1774. *Alceste* is sometimes classified as a *Singspiel,* though the subject is serious and the work features only recitatives rather than spoken dialog. It was the first serious opera to be set in German and was successful enough to be performed as a repertoire piece.

39. Koenigsberger, *Hermann von Helmholtz,* 226.

40. Quoted in an appended biography of Olga and Hermann Helmholtz, written in 1902 by Betty Helmholtz; in Helmholtz, *Letters of Hermann von Helmholtz to His Wife, 1847–1859,* 195–196.

41. In Koenigsberger, *Hermann von Helmholtz,* 226.

42. Helmholtz claimed that he did not usually like lieder at all, though he begrudgingly made an exception for Frau Richelot's performance of Beethoven's "Adelaide": "On Friday evening also I stayed at Richelot's, Frau Richelot sang songs by Beethoven, some of them very well. 'Trocknet nicht Thränen der ewigen Liebe' would be just the thing for you to sing for the *Brunnenkur;* even Adelaide, whom I would like to otherwise not suffer, performed, made it through herself" (Hermann Helmholtz, letter dated July 26, 1855 (no. 37), in Helmholtz, *Letters of Hermann von Helmholtz to His Wife, 1847–1859,* 150).

43. Dahlhaus, *Nineteenth-Century Music,* 22. Dahlhaus argues it was not until romanticism that classicism in music came into existence. In terms of exemplary composers, he explains that Palestrina was considered exemplary for Catholic church music, Bach for Protestant church music, Handel for the oratorio, Gluck for musical tragedy, Mozart for opera buffs, Haydn for the string quartet, Beethoven for the symphony, and Schubert for the lied.

44. See the extract on page 60 (Koenigsberger, *Hermann von Helmholtz,* 232–233).

45. "In the afternoon I changed money, heard a concert by the Austrian regimental band in the coffee house on main, and then went to the theater where I first saw

Dr. Robin, *very well* represented by a young new Devrient, so good that I give up ever playing that part. Then, after a most boring concert of various instrumental and vocal soloists—in which to my surprise Fräulein Bywater from Königsberg delighted the people of Frankfurt through comparatively poorly sung coloratura arias—a most splendid excerpt from a posthumous opera by Mendelssohn, *Lorelei*, was also well represented by Frau Anschutz" (Hermann Helmholtz, letter dated August 18, 1853 (no. 28), in Helmholtz, *Letters of Hermann von Helmholtz to His Wife, 1847–1859*, 106).

46. See the extract on page 65 (Koenigsberger, *Hermann von Helmholtz*, 226).

47. "It was to me as if up to now only your soul would have, with its deeply musical interiority (*Innerlichkeit*), led the harmonies into my understanding. My ears heard only musical figures and my soul heard nothing. Of course it was Mozart's symphony during which I felt this way, one of the finest by him, about which all floated around me in delight. I, as I was there, lonely, abandoned by the beautiful half of my soul, would have just as well heard scales played on the piano" (quoted in an appended biography of Olga and Hermann Helmholtz, written about 1902 by Betty Johannes (Olga Helmholtz's sister), likely in response to a request by Leo Koenigsberger; in Helmholtz, *Letters of Hermann von Helmholtz to His Wife, 1847–1859*, 195).

48. "My room-fellow is the son of a Silesian engineer. . . . He has extraordinary execution on the piano, but only cares for florid pieces and for modern Italian music" (Helmholtz, in Koenigsberger, *Hermann von Helmholtz*, 13–14).

49. "We went to a smoking concert, i.e. to the rehearsal for the great orchestral concerts, where they give certain *soli* that are left out in the concert proper, and which the gentlemen of Utrecht listen to over their wine and cigars. I heard the *Symphonic Preludes* of Liszt, which are effective and extraordinary enough, but hardly beautiful; the *Oberon* Overture was very good, and as a piano solo in between we had the *Variations Sérieuses* of Mendelssohn, in the style of church music, which were very fine, and which I recommend you to study. Donders had been giving public lectures here on Acoustics, so that my book *On the Sensations of Tone* is known to everyone, even to the musicians. O. Jahn could not understand it, but hoped to study it with G., and told me he had had an enthusiastic letter about it from Claus Groth" (Helmholtz, March 14, 1863, letter to his wife, in Koenigsberger, *Hermann von Helmholtz*, 222). Helmholtz was also critical of the "typical insensible style" of Butera's *Atala* and of a bell concert in a Scottish style that he believed could only be appreciated by its inventor (Helmholtz, letters (no. 24 and 25), in Helmholtz, *Letters of Hermann von Helmholtz to His Wife, 1847–1859*, 86, 90).

50. Helmholtz, *On the Sensations of Tone*, 314.

51. Helmholtz, *On the Sensations of Tone*, 314.

52. Helmholtz promoted the cause of just temperament in his 1857 popular lecture "On the Physiological Causes of Harmony in Music," his 1860 lecture "Über

Musikalische Temperatur," and *Tonempfindungen*, and he even referred to "preaching" about unequal temperament to the organ maker M. Cavallie in a letter to his wife (Helmholtz, in Koenigsberger, *Hermann von Helmholtz*, 232).

53. The harmonium, a cheaper, smaller, and more rugged alternative to the pipe organ, was extremely popular in Europe and the United States through the nineteenth and early twentieth centuries. A pumping apparatus vibrates a bank of brass reeds. More sophisticated models (such as the one Helmholtz commissioned) have multiple reed banks and, correspondingly, multiple keyboards. Stops for drones allow the player a certain amount of control over the sound. The instrument's distinct timbre is the result of a double bellows system similar to that of the bagpipes, which allows the harmonium to create a sustained tone.

54. Helmholtz, *On the Sensations of Tone*, 315.

55. This very slight tempering of the fifths is perhaps the reason Helmholtz persisted in describing his commissioned harmonium as "just tuned" even though, by only keeping the thirds perfectly pure, such a system would conventionally be considered a form of mean-tone temperament. Readers familiar with the complexities of tuning systems may find my use of "just temperament" and "just tempered" oxymoronic, but I do so in order to highlight that while Helmholtz claimed that his harmonium was just tuned, it was, because of the slight tempering of the fifths, not. Given Helmholtz's great care with terminology throughout *On the Sensations of Tone*, one can assume that his incorrect description of his harmonium as "just tuned" was deliberate and likely part of his effort to promote just intonation. Alexander Ellis, the renowned English philologist and mathematician, began the English translation, with extensive comments and appendixes, of Helmholtz's text in 1875. He too was quite interested in the musical systems of other cultures as well as the implementation of equal temperament (it is therefore of note and a further indication of Helmholtz's deliberateness that Ellis was willing to translate Helmholtz's description of his harmonium as "just tuned" without any comment). He claimed in a paper read before an 1880 Society of Arts meeting in Britain that though aimed at, equal temperament was only rarely attained (Alexander Ellis, "On the History of Musical Pitch," *Journal of the Society of Arts*, 28 (1880):293–336). For an excellent discussion of the implementation of equal temperament see Ross W. Duffin's *How Equal Temperament Ruined Harmony (and Why You Should Care)* (New York: Norton, 2007). See also Mark Lindley, *Lutes, Viols, and Temperaments* (Cambridge: Cambridge University Press, 1984).

56. Helmholtz, *On the Sensations of Tone*, 319.

57. Helmholtz, *On the Sensations of Tone*, 319.

58. Helmholtz, *On the Sensations of Tone*, 320. Myles Jackson examines the similar concerns of the first half of the nineteenth-century in his chapter, "Physics, Machines, and Musical Pedagogy," *Harmonious Triads*.

59. Helmholtz, *On the Sensations of Tone*, 322–323.

60. Helmholtz, *On the Sensations of Tone*, 322–323.

61. Helmholtz, *On the Sensations of Tone*, 322–323.

62. Helmholtz, *On the Sensations of Tone*, 324–325.

63. Helmholtz, *On the Sensations of Tone*, 324–325.

64. Helmholtz, *On the Sensations of Tone*, 320.

65. Helmholtz, *On the Sensations of Tone*, 327. Such statements can be read as perhaps alluding to the works of Richard Wagner, which were initially criticized as wild and jarring. It is not, however, clear exactly what Helmholtz's opinion of Wagner's music was, though it certainly makes use of an aesthetic very different from the pieces and composers Helmholtz openly admired. His second wife, Anna Helmholtz, maintained a friendly correspondence with Cosima Wagner (Richard Wagner's wife, daughter of Franz Liszt) through the 1890s. Anna Helmholtz attended the Bayreuth festival by herself in 1884, then with Hermann in 1892, and then again alone (after his death) in 1899. See Petra Werner and Angelika Irmscher, *Kunst und Liebe müssen sein: Breife von Anna von Helmholtz an Cosima Wagner 1889 bis 1899* (Bayreuth: Druckhaus Bayreuth, 1993).

66. Helmholtz, *On the Sensations of Tone*, 327.

67. Helmholtz, *On the Sensations of Tone*, 327.

68. Historian David Pantalony's examination of the Parisian workshop of Rudolph Koenig is an excellent example (David Pantalony, *Altered Sensations: Rudolph Koenig's Acoustical Workshop in Nineteenth-Century Paris* (Heidelberg: Springer, 2009)).

69. Helmholtz, "Über Musikalische Temperatur," 74.

70. Helmholtz, "On the Motion of the Strings of a Violin."

71. Helmholtz, *On the Sensations of Tone*, 19, 86.

72. Jackson, "Physics, Machines, and Musical Pedagogy," *Harmonious Triads*.

73. Koenigsberger, *Hermann von Helmholtz*, 232.

74. In Koenigsberger, *Hermann von Helmholtz*, 232–233.

75. Hiebert, "Helmholtz's Musical Acoustics."

76. Helmholtz, *On the Sensations of Tone*, 74–75.

77. Helmholtz, *On the Sensations of Tone*, 75.

78. Myles Jackson includes a nice discussion of Helmholtz's mechanical explanation of the piano and interest in piano makers in *Harmonious Triads*, 272–279.

79. Helmholtz, *On the Sensations of Tone,* 75.

80. Helmholtz, *On the Sensations of Tone,* 76.

81. David Cahan offers a more thorough discussion of the relationship between Helmholtz and the Steinway family as well as of Helmholtz's formal and informal exchange with American science more generally; see his article "Helmholtz in Gilded-Age America: The International Electrical Congress of 1893 and the Relations of Science and Technology," *Annals of Science* 67 (2010): 1–38.

82. Hermann Helmholtz, letter dated June 9, 1871, to Theodore Steinway, Steinway and Sons Catalog, 1872. Cited in Richard K. Lieberman, *Steinway and Sons* (New Haven, CT: Yale University Press, 1995), 61.

83. D. W. Fostle, *The Steinway Saga: An American Dynasty* (New York: Scribner, 1995), 286–287.

84. Lieberman, *Steinway and Sons,* 61.

85. See Hiebert and Hiebert's discussion in "Musical Thought and Practice," 307.

86. These letters were translated and maintained in the Steinway catalog for almost two decades along with endorsements from Lizst, Wagner, and others. One letter, from the Steinways to Helmholtz, is located in the Archiv der Berlin-Brandenburgischen Akademie der Wissenschaften. Two others are on loan to the La Guardia College Archive. Another is in the Steinway Archive and two others are published in the Steinway catalogs. See Hiebert and Hiebert's article "Musical Thought and Practice" for further discussion of these letters.

87. The Steinway sound can be compared to the French Pleyels or Erards, which were known as more round, warm, and sensual (fitting for the music of Frédéric Chopin, who preferred Pleyels), or the Austrian Bösendorfer (one of which Ernst Mach owned), which was considered richer and more full-bodied. This Bösendorfer sound was further reinforced in 1900 with Bösendorfer's introduction of the Imperial Grand Model 290, which included ninety-seven keys, nine more than contemporary keyboards. The additional strings of the Imperial Grand, as well as later models with ninety-two keys, contribute to the richness of the Bösendorfer sound by sympathetically vibrating with played notes (very little music is actually written for the extra keys). Little information exists on the Kaim and Günther piano, the instrument Helmholtz owned before acquiring the Steinway, other than that the German firm was established in 1819, making it one of the oldest of the modern piano-making firms.

88. Hiebert, "Helmholtz's Musical Acoustics," 427. Hiebert refers to Oscar Paul's publication, *Geschichte des Claviers vom Ursprunge bis zu den modernsten Formen dieses Instruments nebst einer Übersicht über Musikalische Abtheilung der Pariser Weltausstellung im Jahre 1867* (Leipzig: A. H. Payne, 1868), 41.

89. Helmholtz, *On the Sensations of Tone*, 6.

90. Helmholtz, *Die Lehre von den Tonempfindungen*, 3rd ed., 406.

91. Helmholtz claimed that everyone already chose natural thirds in harmony. It was in melody that musicians raised on equal temperament would overreach, sharpening the interval (Helmholtz, *Die Lehre von den Tonempfindungen*, 3rd ed., 407).

92. Helmholtz, *Die Lehre von den Tonempfindungen*, 3rd ed., 509.

93. Helmholtz, *Die Lehre von den Tonempfindungen*, 3rd ed., 507–509.

94. Hermann Helmholtz, *On the Sensations of Tone*, trans. Alexander Ellis, 1875 ed., 508.

95. Helmholtz, *Die Lehre von den Tonempfindungen*, 3rd ed., 634.

96. Benjamin Steege has examined the laryngoscopic studies of vocalist and music teacher Emma Seiler, performed in collaboration with Helmholtz in his work on upper partials in vowel sounds. Steege's more general argument about the estrangement of voice from body mediated by the laryngoscopy is discussed in the final chapter of this book (Benjamin Steege, "*Bel Canto* Refracted: Victorian Voices in the Mirror," unpublished manuscript, Columbia University, 2012).

97. Helmholtz, *On the Sensations of Tone*, trans. Alexander Ellis, 1875 ed., 640.

98. Duffin further points out that in 1875, Alexander Ellis, ever the surveyor of pitches, found several professional musicians playing leading notes and major thirds between equal temperament and Pythagorean temperament, far sharper than the pure intervals Helmholtz referred to. Additionally, he postulates that Joachim was accustomed to quartet playing and never adapted to this newer style in which soloists sharpened their leading notes (Duffin, *How Equal Temperament Ruined Harmony*, 122–123, 129).

99. Duffin analyzes several of Joachim's 1903 recordings, measuring the frequency of his intervals, and concludes that he indeed played intervals that were much closer to being pure than equal tempered (Duffin, *How Equal Temperament Ruined Harmony*, 129–133).

100. Notably, Shaw suggested that a physicist construct a tonometer giving a theoretically perfect major scale, "in order that Joachim, Sarasate, Ysaÿe, and Reményi should have an opportunity of hearing how far the four different tone figures which they have made for themselves as major scales differ from the theoretic scale and from one another" (Shaw, *Music in London, 1890–1894*, vol. 2, 276–277).

101. Shaw only begrudgingly accepted Joachim's fame after devoting some very harsh criticisms to his playing. He scathingly described Joachim in an 1890 concert as scraping "away frantically, making a sound after which an attempt to grate a nutmeg effectively on a boot sole would have been as the strain of an Aeolian harp.

The notes which were musical enough to have any discernible pitch at all were mostly out of tune. It was horrible—damnable!" Duffin believes that Shaw's indictment of Joachim's sound was due to Joachim's distinctive tuning. Shaw later concluded that Joachim was a master violinist after all and just employed a different intonation than what he was accustomed to (George Bernard Shaw, *London Music in 1888–89* (New York: Dodd, Mead, and Co., 1937), 332, 350; Shaw, *Music in London, 1890–1894*, vol. 2, 276–277; Duffin, *How Equal Temperament Ruined Harmony*, 124).

102. Beatrix Borchard, "Joseph Joachim," *Grove Music Online*, www.oxfordmusiconline.com.

103. Eduard Hanslick, "Joseph Joachim," in Henry Pleasants, ed. and trans., *Music Criticisms, 1846–1899*, 78–81 (New York: Penguin, [1861] 1950), 80–81.

104. Hanslick, "Joseph Joachim," 80.

105. Joseph Joachim, letter to Clara Schumann, dated December 10, 1855, in Joseph Joachim, *Letters from and to Joseph Joachim*, ed. and trans. Nora Bickley (New York: Macmillan, 1914), 113.

106. Hanslick, "Joseph Joachim," 78.

107. Brahms and Joachim's "Manifesto" was initially published on May 6, 1860, in the *Berliner Musik-Zeitung Echo*. The quote is from the translated version in appendix 1 of David Brodbeck's article, "Brahms and the New German School," in Walter Frisch, ed., *Brahms and His World* (Princeton, NJ: Princeton University Press, 1990), 78.

108. Apparently the composer Carl Friedrich Weitzmann had gotten ahold of the Manifesto and wrote a scathing parody of the letter, run anonymously in the *Neue Zeitschrift* two days prior to the publication of Brahms and Joachim's Manifesto. Brodbeck summarizes the whole affair (and includes Weitzmann's parody as appendix 2) in "Brahms and the New German School," 72–73.

109. Eduard Hanslick, *The Beautiful in Music*, ed. Morris Weitz, trans. Gustav Cohen (New York: Liberal Arts Press, [1854] 1957), 117.

110. Hanslick, *The Beautiful in Music*, 44. Wagner fanned the flames of the antagonism, indeed, made it much more personal with a vicious portrayal of Hanslick in the character of Beckmesser in his 1868 opera *Die Meistersinger*, a draft of which was read in front of Hanslick at Josef Standhartner's home in 1862.

111. Friedrich Nietzsche, "The Case of Wagner," in Friedrich Nietzsche, *Basic Writings of Nietzsche*, trans. Walter Kaufmann (New York: Modern Library, [1888] 1992), 634.

112. Nietzsche, *The Case of Wagner*, 636. In the same passage Nietzsche took a swing at Riemann as well. Musicologist Leslie Blasius explains that Nietzsche was critiquing Riemann's naturalized music theory for overreaching in his descriptive project while simultaneously hesitating in his prescriptive one (Leslie Blasius, "Nietzsche, Riemann, Wagner: When Music Lies," in Suzannah Clark and Alexander

Rehding, eds., *Music Theory and Natural Order from the Renaissance to the Early Twentieth Century*, 93–107 (Cambridge: Cambridge University Press, 2001).

113. Leon Botstein, "Time and Memory in Brahms's Vienna," in Walter Frisch, ed., *Brahms and His World*,3–26 (Princeton, NJ: Princeton University Press, 1990), 7.

114. Botstein, "Time and Memory in Brahms's Vienna," 7.

115. Lawrence Kramer, "Franz Liszt and the Virtuoso Public Sphere: Sight and Sound in the Rise of Mass Entertainment," in Lawrence Kramer, *Musical Meaning: Toward a Critical History*, 68–100 (Berkeley: University of California Press, 2002), 70–75.

116. Kramer, "Franz Liszt and the Virtuoso Public Sphere," 74.

117. Kramer, "Franz Liszt and the Virtuoso Public Sphere," 92.

118. Julia Kursell "Sound Objects," presented at "Sounds of Science, Schall im Labor (1800–1930)" Workshop, Max-Planck-Institut für Wissenschaftsgeschichte, Berlin, October 5–7, 2006.

119. It should be noted that Helmholtz was careful to distinguish between sensation and perception, the latter of which was the brain's processing of the former. When describing a study of sensation, Helmholtz explained that it was difficult to analyze sensations that could not be attached to corresponding differences in an external object, essentially that a psychophysical experiment based on the Fechner-Weber law would not work for a study of the ability to, say, distinguish harmonic overtones. The observer had to train himself to hear them—Helmholtz claimed that he could hear up to the sixteenth upper partial. See Helmholtz, *On the Sensations of Tone*, 49–51.

120. Hermann Helmholtz, "Ueber Combinationstöne," *Poggendorff's Annalen der Physik und Chemie* 99 (1856): 497–540.

121. Gustav Hällstrom, "Von den Combinationstönen," *Annalen*, 2nd series, 24 (1832): 438–466.

122. If the two original sounding tones are close together in frequency, the listener will hear beats. As the difference in the frequency of the sounding tones increases, the frequency of the beats also increases. When the difference between the frequencies of the sounding tones is high enough, the ear would hear the beats instead as an independent tone.

123. R. Steven Turner, "The Ohm-Seebeck Dispute, Hermann von Helmholtz, and the Origins of Physiological Acoustics," *British Journal for the History of Science* 10, no. 34 (1977): 13.

124. Helmholtz, "Ueber Combinationstöne," 273–282.

125. Helmholtz thought that the interference of the original sounding tones' harmonic overtones occurred in the ear canal between the eardrum and ossicles. But

in this case the sound waves of the combination tones were still being created prior to stimulating the auditory nerve and therefore still objectively existed independently of the listener (Helmholtz, "Ueber Combinationstöne," 300).

126. Turner, "The Ohm-Seebeck Dispute, Hermann von Helmholtz, and the Origins of Physiological Acoustics," 14.

127. "In the so-called vestibulum, also, where the nerves expand upon little membranous bags swimming in water, elastic appendages, similar to stiff hairs, have been lately discovered at the ends of the nerves. The anatomical arrangement of these appendages leaves scarcely any room to doubt that they are set into sympathetic vibration by the waves of sound which are conducted through the ear. Now if we venture to conjecture . . . that every such appendage is tuned to a certain tone like the strings of a piano then the recent experiment with a piano shows you that when (and only when) that tone is sounded the corresponding hair-like appendage may vibrate, and the corresponding nerve-fibre experiences a sensation, so that the presence of each single such tone in the midst of a whole confusion of tones must be indicated by the corresponding sensation" (Helmholtz, "On the Physiological Causes of Harmony in Music," 40).

128. Helmholtz, "Ueber die physiologischen Ursachen der musikalischen Harmonie," Bonn lecture presented in 1857, published in *Vorträge und Reden*, vol. 1, 99–155 (Braunschweig: F. Vieweg und Sohn, 1884), 146–147. In his 1869 address, "The Aim and Progress of Physical Science," Helmholtz referred to this earlier discussion of the geistige ear: "Nor need I remind you that the ear conveys to us sounds from without in no way in the ratio of their actual intensity, but strangely resolves them and modifies them, intensifying or weakening them in very different degrees, according to their varieties of pitch" (Helmholtz, "The Aim and Progress of Physical Science," 342).

129. See Gary Hatfield's book, *The Natural and the Normative: Theories of Spatial Perception from Kant to Helmholtz* (Cambridge, MA: MIT Press, 1990), and Timothy Lenoir's article, "The Eye as Mathematician: Clinical Practice, Instrumentation, and Helmholtz's Construction of an Empiricist Theory of Vision," in David Cahan, ed., *Hermann von Helmholtz and the Foundations of Nineteenth-Century Science*, 109–153 (Berkeley: University of California Press, 1993).

130. Helmholtz, "The Aim and Progress of Physical Science," 343.

131. Helmholtz, "Ueber Combinationstöne," 289. There is here a resonance with Lamarckian evolutionary theory and the means by which an individual organism acquired physical traits, through repeated behaviors that over time resulted in a physiological change. Helmholtz certainly held a Lamarckian conception of evolution, describing the structural peculiarities acquired by parents and passed on to their offspring, and acknowledging that a single individual was capable of adapting itself. He understood Darwin's work as providing the law of transmission for this

phenomenon. See Helmholtz, "The Aim and Progress of Physical Science," 337–341. Additionally, Peter Pesic, in his fascinating article, "Helmholtz, Riemann, and the Sirens: Sound, Color, and the Problem of Space," examines Helmholtz's studies of the three-dimensional manifold of color sensation, his work with sirens, and his response to Bernhardt Riemann's criticisms to reveal Helmholtz's analogical thinking connecting the musical scale with space. (Peter Pesic, "Helmholtz, Riemann, and the Sirens: Sound, Color, and the Problem of Space," unpublished manuscript, St. John's College, 2012.)

132. See Hatfield, *The Natural and the Normative*; Hatfield, "Helmholtz and Classicism"; Heidelberger, "Force, Law, and Experiment."

133. Helmholtz, "The Aim and Progress of Physical Science," 325.

134. Helmholtz, *On the Sensations of Tone*, 234–235.

135. Helmholtz, *On the Sensations of Tone*, 235.

136. Helmholtz, *On the Sensations of Tone*, 234–235.

137. Helmholtz, *On the Sensations of Tone*, 234.

138. Hermann Helmholtz, "Über die arabisch-persische Tonleiter," paper presented on May 31, 1862, *Naturh.-med. Verein in Heidelberg* 2:216-217. See also Helmholtz's *On the Sensations of Tone*, 279–285.

139. Helmholtz, *On the Sensations of Tone*, 282–283.

140. Helmholtz suggested that the Alexandrian Greeks likely acquired both tuning traditions as well as fretted and bowed string instruments from the Near East (Helmholtz, *On the Sensations of Tone*, 285).

141. "This reference to the history of music was necessitated by our inability in this case to appeal to observation and experiment for establishing our explanations. . . . If our theory of the modern tonal system is correct it must also suffice to furnish the requisite explanation of the former less perfect stages of development" (Helmholtz, *On the Sensations of Tone*, 249).

Chapter 4

1. Ernst Mach, *The Analysis of the Sensations and the Relation of the Physical to the Psychical*, trans. C. M. Williams (New York: Dover, [1886] 1959), 309. Mach was referring in the last sentence to the work by the founder of Viennese comparative musicology, Richard Wallaschek, *Anfänge der Tonkunst* (Leipzig: J. A. Barth, 1903).

2. Ludwig Karpath wrote: "A close relative has told me a lot about a Vienna coffeehouse which I can no longer remember, the 'Café Elephant,' which was located in a narrow passage between Stephen's Place and the Graben. Every day, scholars,

artists, and doctors of medicine and law would gather together. The regulars (*Stammgaesten*) included such later famous people as Professor Mach, Lynkeus (Popper), a group of Wagner-oriented musicians: Peter Cornelius, Heinrich Proges, the music critic Grad, the composer Goldmark, and many others. People wandered in around 2 p.m. and stayed until 2 in the morning, that is, some were always leaving while others were arriving. Unbroken wit and argument on philosophical, scientific, and artistic matters kept the discussion sharp and stimulating. To a certain extent the young Dozent Ernst Mach presided over the gathering. His profound understanding and reflective manner impressed everyone. According to my relative he was one of the first to occupy himself deeply with the recently published work of Helmholtz on tone perceptions about which he formed many interesting and instructive conclusions" (Letter part of Ernst Anton Lederer's (Mach's grandson's) private collection at his home in Essex Fells, NJ; quoted (and translated) in John Blackmore's *Ernst Mach: His Work, Life, and Influence* (Berkeley: University of California Press, 1972), 23).

3. Ernst Mach, foreword (written in 1906) to Eduard Kulke's *Kritik der Philosophie des Schönen* (Leipzig: Deutsche Verlagactiengesellschaft, 1906), x–xi.

4. Kulke, foreword (written in 1896) to *Kritik der Philosophie des Schönen* (Leipzig: Deutsche Verlagactiengesellschaft, 1906), vii–viii.

5. "To prove that Wagner's music could be beautiful, I was not capable; but that it pleased me, that I knew. And the gentlemen and ladies of aesthetic education superior to me, what did they have on me? They believed to be able to prove that Wagner's music could not be beautiful; but could they also cause it to not please me? I had the feeling that all of the music scholars of the entire world, if they met for such a purpose, could not have brought this about. How did they even begin to know better what occurred in me, when I knew it!" (Kulke, foreword (written in 1896) to *Kritik der Philosophie des Schönen*, viii).

6. Ernst Mach, *Einleitung in die Helmholtz'sche Musiktheorie* (Graz: Leuschner & Lubensky, 1866), v, 4.

7. See David C. Large and William Weber, eds., *Wagnerism in European Culture and Politics* (Ithaca, NY: Cornell University Press, 1984); Celia Applegate and Pamela Potter, eds., *Music and German National Identity* (Chicago: University of Chicago Press, 2002).

8. *Do* try this on a piano or guitar. If possible have someone else play the chords so that you are free to concentrate on listening.

9. Mach presented this example in his 1865 lecture "Die Erklärung der Harmonie," in Ernst Mach, *Zwei populäre Vorlesungen über musikalische Akustik* (Graz: Leuschner & Lubensky, 1865), and in his article of the same year, "Bemerkungen über die Accommodation des Ohres."

10. See the discussion of Helmholtz's theory of signs in chapter 3.

11. A special thank you to Peter Pesic for bringing Bernhard Riemann's work on sign theory and accommodation to my attention. The last major (though uncompleted and unpublished) work by mathematician Bernhard Riemann, published posthumously in 1866, "Mechanik des Ohres," presents a criticism of Helmholtz's study of sound sensation as "synthetic" rather than "analytic," in part because he did not think that the sign theory could be directly extended to hearing. If this were the case, the tension on the tensor tympani muscle would have had to be constantly changing. Like Mach, Riemann instead believed that the distinction between the alert ear and the nonalert ear (by which I think we can assume he is referring to the phenomenon of accommodation in hearing) was "dependent on whether or not the foot of the stirrup is pressed lightly against the inner ear fluid by tension of *M. tensor tympani*, so that the pressure at the inner ear fluid is slightly greater than that of the air in the tympanic cavity" (Bernhard Riemann, "The Mechanism of the Ear," *Fusion*, 6, no. 3 ([1866] 1984): 31–38 (quote on 37)). See also Peter Pesic's article on both Riemann's work and Helmholtz's response to it: "Helmholtz, Riemann, and the Sirens: Sound, Color, and the Problem of Space."

12. Michael Hagner, "Toward a History of Attention in Culture and Science," *MLN* 118, no. 3 (2003): 670–687.

13. Hagner, "Toward a History of Attention in Culture and Science," 680.

14. Hagner, "Toward a History of Attention in Culture and Science," 681.

15. Ernst Mach, "Zur Theorie des Gehörorgans," *Sitzungsberichte der kaiserlichen Akademie der Wissenschaften* 48, no. 2 (1863): 283–300.

16. Mach's physical proof consisted of mathematically demonstrating that the eardrum, ossicles, and fluid of the labyrinth all vibrate to regulate the transmission of sound waves in two significant ways: even absorption and quick dissipation of the initial state of the sound waves. The derivation itself consisted of setting up the following second-order equation in which the first coefficient ($-p^2$) is the restoring force of the system, the second coefficient ($-2(b+\beta)$) is the viscosity of the fluid of the labyrinth, and $2abq \cos (qt+\beta)$ is the speed of the air particles in which $2b$ is the elasticity coefficient: $dx^2/dt^2 = -p^2x - 2(b+\beta)(dx/dt) + 2abq \cos(qt+\tau)$. Mach then integrated the equation, solving for x and showing, essentially, the inverse relationship of air particle speed and elastic resistance of the ear components. Thus, the restoring force of the vibrating ear bones and the viscosity of the fluid of the labyrinth both equalized the absorption of sound waves of varying frequency (tones of different pitch) and also dampened resonance (harmonic overtones of the original sound wave), allowing for the transmission of a quick succession of tones. Mach's belief that all of these components—the eardrum, the ossicles, and the fluid—played a role in sound-wave transmission was rooted in earlier work by Savart and Seebeck. Mach further posited the possibility that different types of transmission occur in the ear depending on the wavelength of sound waves. He suggested that the

eardrum, ossicles, and fluid of the labyrinth all vibrate together to transmit lower tones but vibrate separately to transmit higher tones. Although this hypothesis did not provide a mechanism of accommodation, it was at least a physical explanation in which different pitches were treated differently in the ear (Mach, "Zur Theorie des Gehörorgans," 285–287).

17. Mach, "Zur Theorie des Gehörorgans," 289.

18. Mach later revisited this experiment and concluded that it was more likely that the pinching created reflected waves in the tube and that the interference of these waves with the original ones was what in fact led to the weakening of the volume of the sung tone in his ears (Ernst Mach, "Über einige der physiologischen Akustik angehörige Erscheinungen," *Sitzungsberichte der kaiserlichen Akademie der Wissenschaften* 50 (1864): 345).

19. Mach, "Zur Theorie des Gehörorgans," 297.

20. Mach, "Bemerkungen über die Accommodation des Ohres," 345.

21. Mach, "Bemerkungen über die Accommodation des Ohres," 345. Mach was referring to Politzer's 1864 article with Trötsch in *Archiv für Ohrenheilkunde*.

22. Mach, "Zur Theorie des Gehörorgans," 299–300.

23. Ernst Mach, "Die Cortischen Fasern des Ohres," reprinted in Ernst Mach, *Populär-Wissenschaftlische Vorlesungen* (Leipzig: J. A. Barth, 1896), 25.

24. Mach gives other similes to illustrate his point: a dog's response to "Phylax," his name, and utter indifference to such other heroic names as "Hercules" or "Plato," or the unified throb of two hearts in love ("Die Cortischen Fasern des Ohres," 23–24). This theory of sympathetic vibration was actually the main target of Mach's eventual criticism of Helmholtz. While Mach maintained Müller's one-to-one correlation of sound-wave vibration rate to nervous end organ, he discarded Müller's attribution of specific energies to each of these nervous end organs (Helmholtz had adopted both). Mach would eventually substitute an assumption of only two energies that were excited in proportion to each other by different rates of vibration. It was these ratios of vibration rates that, regardless of pitch, rendered, in combination with attention, melodic and harmonic combinations of tones intelligible. By 1885 Mach distinguished his hypothesis as psychophysical, the product of physical, physiological, and psychological analysis, in contrast to Helmholtz's efforts to reduce tone sensation to a strictly physical foundation.

25. Mach, "Die Cortischen Fasern des Ohres," 27.

26. Mach, "Die Cortischen Fasern des Ohres," 28.

27. "Es gibt eine Kunst des Komponisten, die Aufmerksamkeit des Hörens zu leiten. Es gibt aber ebenso wohl eine Kunst des Hörens, die auch nicht jedermanns Sache

ist" (Ernst Mach, "Die Erklärung der Harmonie," reprinted in Ernst Mach, *Populär-Wissenschaftliche Vorlesungen* (Leipzig: J. A. Barth, 1896), 37).

28. Mach developed a fuller, generally celebratory, discussion of Helmholtz's theories in his *Einleitung in die Helmholtz'sche Musiktheorie*.

29. Ernst Mach and Johann Kessel, "Versuche über die Accommodation des Ohres," *Sitzungsberichte der kaiserlichen Akademie der Wissenschaften* 66 (1872): 337–343.

30. The vibration microscope, first developed in 1855 by Jules Lissajous, who would later become the scientific consultant on Napoléon III's commission to establish a standard pitch, consisted of a small lens attached by a small arm to a vibrating object, traditionally a tuning fork. Objects (a light beam or stylus mark) viewed through this lens would thus appear to vibrate, tracing out a one-dimensional path of oscillation. If the object viewed through the lens (often a mark on another tuning fork) was itself also vibrating perpendicular to the plane of the primary vibrating tuning fork, a two-dimensional curve would appear to be traced out. These Lissajous curves or figures could then be analyzed relative to the known frequency of the primary tuning fork. Rudolph Koenig produced the best and most sought after Lissajous apparatuses among many world-class precision acoustical instruments. Helmholtz later improved on Lissajous' design with the introduction of the electric motor to drive the vibration of the tuning fork in order to analyze the motion of violin strings. In terms of his accommodation experiments, it appears that Mach viewed the oscillating stirrup thread through the vibrating lens of the vibration microscope and then projected the two-dimensional path onto a screen. See Steven Turner's article on the development of the Lissajous apparatus, "Demonstrating Harmony: Some of the Many Devices Used to Produce Lissajous Curves before the Oscilloscope," *Rittenhaus* 11 (2): 33–51.

31. Mach and Kessel, "Versuche über die Accommodation des Ohres."

32. Ernst Mach and Johann Kessel, "Die Funktion der Trommelhöhle und der Tuba Eustachii," *Sitzungsberichte der kaiserlichen Akademie der Wissenschaften* 66 (1872): 329–336.

33. In a later experiment on a prepared, nonliving ear, they found that the ossicles only vibrate—that sound waves were only transmitted—when the Eustachian tube was closed off (Mach and Kessel, "Die Funktion der Trommelhöhle und der Tuba Eustachii," 332).

34. Mach and Kessel, "Versuche über die Accommodation des Ohres," 342.

35. Ernst Mach to Eduard Kulke, undated (likely early 1860s) letter (no. 1), Ernst Mach Papers, Dibner Library of the History of Science and Technology, Smithsonian Institution Special Collections, Washington, DC.

36. Mach, *Einleitung in die Helmholtz'sche Musiktheorie*, vi–vii.

37. Mach, *Einleitung in die Helmholtz'sche Musiktheorie*, vii.

38. Mach, *Einleitung in die Helmholtz'sche Musiktheorie*, 3.

39. Mach, *Einleitung in die Helmholtz'sche Musiktheorie*, 4.

40. Review of Mach's *Einleitung in die Helmholtz'sche Musiktheorie, Leipziger Allgemeine Musikalische Zeitung* 7 (1867): 58.

41. "Zur Theorie der Musik: Die Physiker und die Musiker," *Leipziger Allgemeine Musikalische Zeitung* 21 (1867): 165–169.

42. "Zur Theorie der Musik," 167.

43. Hermann Mendel, ed., *Musikalisches Conversations-Lexikon*, vol. 10 (Berlin: R. Oppenheim, 1878), 246.

44. "Mach, Ernst," in Hermann Mendel, ed., *Musikalisches Conversations-Lexikon*, vol. 7 (Berlin: R. Oppenheim, 1875), 4.

45. See Jacques Le Rider, *Modernity and Crises of Identity: Culture and Society in Fin-de-Siècle Vienna*, trans. Rosemary Morris (New York: Continuum, 1993); Carl Schorske, *Fin-de-Siècle Vienna: Politics and Culture* (New York: Vintage Books, 1981).

46. See Edward Said's *Orientalism* (Pantheon Books, 1978) for a general but thorough discussion of the West's construction of "the Orient" as "other." See also Philip Bohlman's discussion of race and placelessness in music in "The Remembrance of Things Past: Music, Race, and the End of History in Modern Europe," in Ronald Radano and Philip V. Bohlman, eds., *Music and the Racial Imagination*, 644–676 (Chicago: University of Chicago Press, 2000). Musicologist Julie Brown highlights a related strain of racialism in music bound to the rising late nineteenth-century interest in *Volkmusik*: a complicated conception of hybridization in music. Certain intermixtures associated with the trope of the noble savage could be beneficial. Others, associated with degenerate musical traditions, were far more threatening. She traces Béla Bartók's attitudes toward Gypsy music, shifting from negative to positive rhetoric (Julie Brown, "Bartók, the Gypsies, and Hybridity in Music," in Georgina Born and David Hesmondhalgh, eds., *Western Music and Its Others: Difference, Representation, and Appropriation in Music*, 119–142 (Berkeley: University of California Press, 2000)).

47. The original version was published in the *Neue Zeitschrift für Musik* and therefore especially stung the Leipzig music world, which had lost Mendelssohn only three years prior. Beyond an outcry from the faculty of the Leipzig Conservatory, however, there was little reaction to the piece. In the second, 1869 version, Wagner additionally singled Hanslick out for attack, which of course fanned the flames of their ongoing conflict.

48. The classic example is of the giraffe: an individual giraffe, motivated by a change in environment, "wants" the juicy leaves higher up on a tree and repeatedly stretches

its neck to reach these leaves. Through this increased use it undergoes a physiological change, developing a longer neck. Decreased use could, of course, lead to the loss of a physiological trait.

49. Peter Bowler, *Evolution: The History of an Idea*, 3rd ed. (Berkeley: University of California Press, 2003), 93.

50. The term *neo-Lamarckism* was in fact popularized in the 1890s by George John Romanes's description of an increasingly polarized difference of opinion between the neo-Darwinists and neo-Lamarckians (Bowler, *Evolution*, 236).

51. Bowler, *Evolution,* 238.

52. Ewald Hering, "On Memory as a General Function of Organized Matter," in Samuel Butler, ed., *Unconscious Memory,* 2nd ed., 80–83 (New York: Dutton, 1911).

53. In his 1860 book, *Elemente der Psychophysik,* Gustav Fechner presented a monistic understanding of the world in which psychical and physical experiences were two different perspectives of the same event, two sides of the same reality. He presented the example of a circle differing depending on whether one stood inside or outside. It was impossible, when standing on the plane of the circle, to simultaneously experience both the convex and concave sides (Gustav Fechner, *Elemente der Psychophysik,* vol. 1 (Amsterdam: E. J. Bonset, [1860] 1864), 2–3).

54. Hering, "On Memory as a General Function of Organized Matter," 80.

55. Hering, "On Memory as a General Function of Organized Matter," 83.

56. Hering, "On Memory as a General Function of Organized Matter," 83.

57. Ernst Mach, "Die Ökonomische Natur der physikalischen Forschung," *Almanach der Kaiserlichen Akademie der Wissenschaft* 32 (1882): 293-319 (quote on 298).

58. Mach, "Die Ökonomische Natur der physikalischen Forschung," 298–299.

59. Mach, "Die Ökonomische Natur der physikalischen Forschung," 304. Notice that according to Mach there was a deliberateness required to economically convey knowledge to the next generation. It was not necessarily materially rooted in the memory of a nerve substance of the organism as Hering argued.

60. Ernst Mach, "Why Has Man Two Eyes?," in Ernst Mach, *Popular Scientific Lectures*, trans. Thomas McCormack (Chicago: Open Court, 1898), 82.

61. Ernst Mach, "Ueber die physikalische Bedeutung der Gesetze der Symmetrie," *Lotos. Zeitschrift für Naturwissenschaften* 21 (1871): 139–147.

62. Those with extensive musical training may wonder if Mach made any mention of invertible counterpoint here. He did not.

63. Mach, "Ueber die physikalische Bedeutung der Gesetze der Symmetrie," 145–147.

64. This was, Mach explained, why symmetry was still appreciated by those that had, for example, lost an eye (Mach, "Ueber die physikalische Bedeutung der Gesetze der Symmetrie," 144).

65. Mach, "Ueber die physikalische Bedeutung der Gesetze der Symmetrie," 144.

66. Letter from Ernst Mach to Eduard Kulke, May 30, 1872 (no. 4), Ernst Mach Papers, Dibner Library of the History of Science and Technology, Smithsonian Institution Special Collections, Washington, DC.

67. Letter from Eduard Kulke to Ernst Mach, October 26, 1872, Ernst Mach Nachlass, Deutsches Museum Archives, Munich.

68. Fifty years later, folklorist and composer Béla Bartók maintained a similarly imprecise but also similarly evolutionary and psychophysiological position, explaining that "peasant music is the outcome of changes wrought by a natural unconscious; it is impulsively created by a community of men who have had no schooling; it is as much a natural product as are the various forms of animal and vegetable life" (Béla Bartók, *The Hungarian Folk Song*, ed. Benjamin Suchoff, trans. M. D. Calvocoressi (Albany: State University of New York Press, 1981), 3).

69. Eduard Kulke, *Über die Umbildung der Melodie: Ein Beitrag zur Entwickelungslehre* (Prague: J. G. Calve'sche K. K. Hof- und Univ.-Buchhandlung, 1884). Kulke had earlier referred to the Andante theme of Beethoven's C-Minor Symphony as another fine example of the application of Darwinian evolution to the arts in his October 26, 1872 letter to Mach.

70. Kulke, *Über die Umbildung der Melodie*, 19–20. Kulke pointed to the lukewarm acceptance of modern music as an example of the gradual nature of this process.

71. Kulke, *Über die Umbildung der Melodie*, 5.

72. This reduction of melody to simpler, purer structures bears the stamp of both the formalist analytical and pedagogical technique of Marx discussed in chapter 2 as well as an older Goethean analytical practice of reducing melodies to their ur- or skeletal form.

73. Kulke, *Über die Umbildung der Melodie*, 15.

74. Heinrich Schenker, "Das Hören in der Musik," *Neue Revue* 5 (July 25, 1894): 115–121.

75. Schenker explained that the acoustician may very well be able to describe the overtone series with flawless precision, but would be on "slippery ground" in attempting to apply the knowledge to music or the musician because his reason was "neither enlightened nor corrected by any kind of artistic intuition" (Heinrich Schenker, *Harmony*, ed. Oswald Jonas, trans. Elisabeth Borgese (Chicago: University of Chicago Press, [1906] 1954), 21). Also, Nicholas Cook describes the unpublished manuscript

of "Das Tonsystem" in which Schenker names Helmholtz specifically (see "Foundations of the Schenker Project," in Nicholas Cook, *The Schenker Project: Culture, Race and Music Theory in Fin-de-Siècle Vienna* (Oxford: Oxford University Press, 2007), 77n81).

76. Cook, *The Schenker Project*, 184–186.

77. Cook, *The Schenker Project*, 48–52.

78. Cook, *The Schenker Project*, 54.

79. Heinrich Schenker, "Der Geist der musikalischen Technik," *Musikalisches Wochenblatt* 26 (1895): 245–246, 257–259, 273–274, 279–280, 309–310, 325–326; reprinted, Leipzig: E. W. Fritsch, 1895.

80. Postcard from Mach to Schenker, dated December 2, 1896. Held in the Oswald Jonas Memorial Collection, Special Collections, University of California, Riverside Libraries, box 12, folder 47, and available online at http://mt.ccnmtl.columbia.edu/schenker/correspondence/postcard/oj_1247_12296.html.

81. Schenker, *Harmony*, 5–6.

82. Schenker, *Harmony*, 6–7.

83. Schenker, *Harmony*, 13.

84. All art besides music was, according to Schenker, based on the association of ideas, reflected from nature and reality. But there was no clear association of ideas between music and nature. This was why he believed that the music of what he termed "primitive" people never developed into art and remained instead in a rudimentary stage, supporting his hierarchy of music cultures (Schenker, *Harmony*, 3–5).

85. Not much can be made of this but it is worth noting: in a lab notebook entry dated February 17, 1877, along with diagrams of the overtone series for major triad chords, mentions of tuning forks (*Stimmgabeln*), and the hypothesis of combination tones, Mach wrote "Chopin. Wagner. Beethoven" (Ernst Mach Nachlass, Notizbuch 514, Deutsches Museum, Munich).

86. Dieter Borchmeyer, *Drama and the World of Richard Wagner* (Princeton, NJ: Princeton University Press, 2003), 101.

87. Thomas Mann has pointed out that, while Wagner could strike a popular German note for the purposes of characterization, this was not the wellspring from which his tone poems sprang, concluding that "however authentic and potent it may have been, Wagner's German-ness was refracted and fragmented in a modern way; it was decorative, analytical, and intellectual, hence his fascination, his inability to make a cosmopolitan, planetary impact" [Thomas Mann, "Betrachtungen eines Unpolitischen," in Thomas Mann, ed., *Gesammelte Werke*, vol. 12, 76–77 (Frankfurt: S. Fischer, 1974) quoted in Deiter Borchmeyer, *Drama and the World of Richard Wagner* (Princeton: Princeton University Press, 2003), 102].

88. This conversation was described by Henry C. Lunn, "Free and Cheap Concerts for the People," *Musical Times* 20 (June 1) 1879: 306–307.

89. This Schopenhauerian listening (ear?) of Kulke's was a bit unique and criticized as such. Certainly Kulke's intepretation of Wagner elicited the sarcastic wrath of a British music critic relaying this entire episode in 1879, to say nothing of the individual with whom Kulke was arguing, the "sometimes a singing dragon is just a singing dragon" guy.

90. Mach, *The Analysis of Sensations and the Relation of the Physical to the Psychical,* 30.

91. Ernst Mach, *Die Analyse der Empfindungen und das Verhältnis des Physischen zum Psychischen,* 8th ed. (Jena: G. Fischer, [1886] 1902), 251.

92. Mach, *Die Analyse der Empfindungen und das Verhältnis des Physischen zum Psychischen,* 251.

93. Mach, *The Analysis of the Sensations and the Relation of the Physical to the Psychical,* 309. Mach was referring in the last sentence to the work by the founder of Viennese comparative musicology, Richard Wallaschek, *Anfänge der Tonkunst* (Leipzig: J. A. Barth, 1903).

94. Mach, *Die Analyse der Empfindungen und das Verhältnis des Physischen zum Psychischen,* 24.

95. Kulke, *Kritik der Philosophie des Schönen,* 341–342.

Chapter 5

1. Edwin Boring, "The Psychology of Controversy," *Psychological Review* 36 (1929): 97–121.

2. "The Musical Scale is not one, not 'natural,' nor even founded necessarily on the laws of the constitution of musical sound, so beautifully worked out by Helmholtz, but very diverse, very artificial, and very capricious" (Alexander Ellis, "On the Musical Scales of Various Nations," *Journal of the Society of Arts* 33 (1885): 485–527 (especially 527)).

3. Ellis, "On the Musical Scales of Various Nations," 527.

4. Carl Stumpf, for one, presented a summary review of Ellis's article the following year (Carl Stumpf, "Alexander J. Ellis, on the Musical Scales of Various Nations," *Vierteljahrsschrift für Musikwissenschaft* 2 (1886): 511–524).

5. See Carl Dahlhaus, *Nineteenth-Century Music,* trans. J. Bradford Robinson (Berkeley: University of California Press, 1989); Annegret Fauser, *Musical Encounters at the 1889 Paris World's Fair* (Rochester, NY: University of Rochester Press, 2005); Alex Ross, *The Rest Is Noise: Listening to the Twentieth Century* (New York: Farrar, Straus and Giroux, 2007).

6. Fauser, *Musical Encounters at the 1889 Paris World's Fair*, 165–214.

7. Julie Brown, "Bartók, the Gypsies, and Hybridity in Music," in Georgina Born and David Hesmondhalgh, eds., *Western Music and Its Others: Difference, Representation, and Appropriation in Music*, 119–142 (Berkeley: University of California Press, 2000).

8. Elfrieda Hiebert and Erwin Hiebert discuss Schoenberg's interest in Helmholtz and his efforts to incorporate Helmholtz's theories into his compositions in their article "Musical Thought and Practice: Links to Helmholtz's *Tonempfindungen*," in Lorenz Krüger, ed., *Universalgenie Helmholtz: Rückblick nach 100 Jahren* 295–314 (Berlin: Akademie Verlag, 1994). See also Julia Kursell, "Experiments on Tone color in Music and Acoustics: Helmholtz, Schoenberg, and *Klangfarben melodie*," in *Music, Sound, and the Laboratory during the Nineteenth and Twentieth Centuries, Osiris* vol. 28, forthcoming 2013.

9. Arnold Schoenberg, "Composition with Twelve Tones (I)" in Leonard Stein, ed., and Leo Black, trans., *Style and Idea: Selected Writings of Arnold Schoenberg* (Berkeley: University of California Press, [1941] 1985), 217.

10. Anton Webern, *The Path to the New Music*, ed. Willi Reich (Bryn Mawr, PA: Theodore Presser, [1960] 1963), 48.

11. Kurt Danziger, *Constructing the Subject: Historical Origins of Psychological Research* (Cambridge: Cambridge University Press, 1990). See also M. Norton Wise, "How Do Sums Count?", in Lorenz Krüger, Lorraine Daston, and Michael Heidelberger, eds., *The Probabilistic Revolution*, vol. 1, 395–425 (Cambridge, MA: MIT Press, 1987); William Woodward, "Wundt's Program for the New Psychology: Vicissitudes of Experiment, Theory, and System," in William Woodward and Mitchell G. Ash, eds., *The Problematic Science: Psychology in Nineteenth-Century Thought*, 167–197 (New York: Praeger, 1982).

12. Mitchell Ash, *Gestalt Psychology in German Culture, 1870–1967: Holism and the Quest for Objectivity* (Cambridge: Cambridge University Press, 1995); Woodward and Ash, *The Problematic Science*; Adrian Brock, ed., *Internationalizing the History of Psychology* (New York: New York University Press, 2006); Allan R. Buss, ed., *Psychology in Social Context* (New York: Irvington, 1979; Bruno Nettl and Philip V. Bohlman, eds., *Comparative Musicology and Anthropology of Music: Essays on the History of Ethnomusicology* (Chicago: University of Chicago Press, 1991).

13. Boring, "The Psychology of Controversy," 110.

14. Boring, "The Psychology of Controversy," 99.

15. Boring, "The Psychology of Controversy," 109–111.

16. Carl Stumpf, "Autobiography of Carl Stumpf," in Carl Murchison, ed., *History of Psychology in Autobiography*, vol. 1, 389–441 (Worcester, MA: Clark University Press, [1924] 1930), 396.

17. Carl Lorenz, "Untersuchungen über die Auffassung von Tondistanzen," *Philosophische Studien* 6 (1890): 26–103.

18. Carl Stumpf, "W. Wundt, Grundzüge der physiologischen Psychologie; E. Luft, Über die Unterschiedsempfindlichkeit für Tonhöhern," *Vierteljahrsschrift für Musikwissenschaft* 4 (1888): 545.

19. Carl Stumpf, "Über Vergleichungen von Tondistanzen," *Zeitschrift für Psychologie und Physiologie der Sinnesorgane* 1 (1890): 431–436.

20. Stumpf, "Über Vergleichungen von Tondistanzen," 433.

21. Stumpf, "Über Vergleichungen von Tondistanzen," 453.

22. Stumpf, "Über Vergleichungen von Tondistanzen," 423–426.

23. Stumpf, "Über Vergleichungen von Tondistanzen," 457.

24. Stumpf, "Über Vergleichungen von Tondistanzen," 456.

25. Stumpf, "Über Vergleichungen von Tondistanzen," 450.

26. Wilhelm Wundt, "Ueber Vergleichungen von Tondistanzen," *Philosophische Studien* 6 (1891): 605–640 (especially 609, 632).

27. Stumpf, "W. Wundt, Grundzüge der physiologischen Psychologie; E. Luft, Über die Unterschiedsempfindlichkeit für Tonhöhern," 545.

28. Stumpf, "Über Vergleichungen von Tondistanzen," 420.

29. Carl Stumpf, "Wundt's Antikritik," *Zeitschrift für Psychologie und Physiologie der Sinnesorgane* 2 (1891): 291–292.

30. Stumpf, "Über Vergleichungen von Tondistanzen," 456.

31. Here Stumpf was likely referring to his earlier research, discussed in his *Tonpsychologie*, vol. 1 (Leipzig: Verlag von Hirzel, 1883).

32. Stumpf, "Über Vergleichungen von Tondistanzen," 457.

33. Stumpf, "W. Wundt, Grundzüge der physiologischen Psychologie; E. Luft, Über die Unterschiedsempfindlichkeit für Tonhöhern," 548.

34. For example, Stumpf took issue with such tiny details as Wundt's characterization of the E-flat minor chord at the beginning of Wagner' *Das Rheingold*, the first of his operatic cycle, *Der Ring des Niebelungen*. This illustrates just how bitter the controversy had become. Stumpf was fussing over a detail irrelevant to Wundt's larger point. Apparently on page 79 of the third edition of *Grundzüge der physiologischen Psychologie*, Wundt had described the sustained E-flat minor chord at the beginning of Wagner's *Nibelungen* as the border of the organ note. But, Stumpf explained, it was no such thing, and was in fact the contrary. The border of the organ note is the *change* (*die Veränderung*) in harmony, the changing of the notes, not the sustained chord (Stumpf, "W. Wundt, Grundzüge der physiologischen Psychologie; E. Luft, Über die Unterschiedsempfindlichkeit für Tonhöhern," 450, 542).

35. Stumpf, "W. Wundt, Grundzüge der physiologischen Psychologie; E. Luft, Über die Unterschiedsempfindlichkeit für Tonhöhern," 546.

36. Stumpf, "Wundt's Antikritik," 276.

37. Wundt, "Ueber Vergleichungen von Tondistanzen," 616–617.

38. Wundt, "Ueber Vergleichungen von Tondistanzen," 624.

39. Wundt, "Ueber Vergleichungen von Tondistanzen," 617.

40. Stumpf, "Über Vergleichungen von Tondistanzen," 456.

41. Stumpf, "Autobiography of Carl Stumpf," 401.

42. Boring, "The Psychology of Controversy," 116–117.

43. Stumpf, *Tonpsychologie*, vol. 1, 247.

44. Letter from Carl Stumpf to Joseph Joachim, dated March 5, 1907, held in the Handschriftenarchiv und Photokopiensammlung of the Bibliotek des Staatlichen Instituts für Musikforschung, Berlin.

45. Carl Stumpf, "Phonographirte Indianermelodien," *Vierteljahrsschrift für Musikwissenschaft* 8 (1892): 127–144; Carl Stumpf, "Tonsystem und Musik der Siamesen," *Beiträge zur Akustik und Musikwissenschaft* 3 (1901): 69–146.

46. Carl Stumpf, *Die Anfänge der Musik* (Leipzig: J. A. Barth, 1911), 95.

47. Eduard Hanslick, "Ein Brief über die 'Clavierseuche,'" *Suite. Aufsätze über Musik und Musiker* (Techen: K. Prochaska, 1884).

48. Friedrich Nietzsche, *On the Genealogy of Morality: A Polemic* (Indianapolis: Hackett [1887] 1998); Max Nordau, *Degeneration*, reprinted from English-language ed. published in 1895, translated from 2nd ed. of the German work (Lincoln: University of Nebraska Press, [1892] 1993).

49. Myles Jackson, "Physics, Machines, and Musical Pedagogy," in Myles Jackson, *Harmonious Triads: Physicists, Musicians, and Instrument Makers in Nineteenth-Century Germany*, 231–279 (Cambridge, MA: MIT Press, 2006); James Kennaway, "From Sensibility to Pathology: The Origins of the Idea of Nervous Music around 1800," *Journal of the History of Medicine and Allied Sciences* 65 (2010): 396–426; James Kennaway, *Bad Vibrations: The History of the Idea of Music as a Cause of Disease* (Farnham, Survey: Ashgate, 2012).

50. Friedrich Nietzsche, *Der Fall Wagner: Ein Musikanten-Problem* (Leipzig: Verlag von C. G. Neumann, 1888).

51. Alberto Cambrosio, Daniel Jacobi, and Peter Keating, "Ehrlich's 'Beautiful Pictures' and the Controversial Beginnings of Immunological Imagery," *Isis* 84 (1993): 662–699; Harry Collins, *Changing Order: Replication and Induction in Scientific Practice*

(Chicago: University of Chicago Press, 1992); Peter Galison, *Image and Logic: A Material Culture of Microphysics* (Chicago: University of Chicago Press, 1997); Bruno Latour and Steve Woolgar, *Laboratory Life: The Construction of Scientific Facts* (Princeton, NJ: Princeton University Press, 1986); Timothy Lenoir, "Inscription Practices and Materialities of Communication," in Timothy Lenoir, ed., *Inscribing Scientific Texts and the Materiality of Communication*, 1–19 (Stanford, CA: Stanford University Press, 1998); Michael Lynch, *Art and Artifact in Laboratory Science: A Study of Shop Work and Shop Talk in a Research Laboratory* (London: Routledge & Kegan Paul, 1985); Michael Lynch, "The Externalized Retina: Selection and Mathematization in the Visual Documentation of Objects in the Life Sciences," in Michael Lynch and Steve Woolgar, eds., *Representation in Scientific Practice*, 153–186 (Cambridge, MA: MIT Press, 1988); Martin Rudwick, "The Emergence of a Visual Language for a Geological Science, 1760–1840," *History of Science* 14 (1976): 149–195; Steve Shapin and Simon Schaffer, *Leviathan and the Air-Pump: Hobbes, Boyle, and the Experimental Life* (Princeton, NJ: Princeton University Press, 1985).

52. Karin Bijsterveld, "The Diabolical Symphony of the Mechanical Age: Technology and Symbolism of Sound in European and North American Noise Abatement Campaigns, 1900–40," *Social Studies of Science* 31 (2001): 37–70; Karin Bijsterveld, "Listening to Machines: Industrial Noise, Hearing Loss and the Cultural Meaning of Sound," *Interdisciplinary Science Reviews* 31 (2006): 323–377; Robert Brain, "Standards and Semiotics," in Timothy Lenoir, ed., *Inscribing Science: Scientific Texts and the Materiality of Communication*, 249–284 (Stanford, CA: Stanford University Press, 1998); Cyrus Mody, "The Sounds of Science: Listening to Laboratory Practice," *Science, Technology, & Human Values* 30, no. 2 (Spring 2005): 175–198; Trevor Pinch and Frank Trocco, *Analog Days: The Invention and Impact of the Moog Synthesizer* (Cambridge, MA: Harvard University Press, 2002); Emily Thompson, "Dead Rooms and Live Wires: Harvard, Hollywood, and the Deconstruction of Architectural Acoustics, 1900–1930," *Isis* 88 (1997): 597–626; Emily Thompson, "Listening to/for Modernity: Architectural Acoustics and the Development of Modern Spaces in America," in Peter Galison and Emily Thompson, eds., *The Architecture of Science*. 352–380 (Cambridge, MA: MIT Press, 1999); Adelheid Voskul, "Humans, Machines, and Conversations: An Ethnographic Study in the Making of Automatic Speech Recognition Technologies," *Social Studies of Science* 34 (2004): 393–421.

53. Ben Steege, "*Bel Canto* Refracted: Victorian Voices in the Mirror," unpublished manuscript, Columbia University, 2012, 17.

54. Mody, "The Sounds of Science."

55. Philip Bohlman, "Composing the Cantorate: Westernizing Europe's Other within," in Georgina Born and David Hesmondhalgh, eds., *Western Music and Its Others* (Berkeley: University of California Press, 2000), 188.

56. Bohlman, "Composing the Cantorate," 190–191.

57. See Jacques Le Rider, *Modernity and Crises of Identity: Culture and Society in Fin-de-Siècle Vienna*, trans. Rosemary Morris (New York: Continuum, 1993); David Myers, *Resisting History: Historicism and Its Discontents in German-Jewish Thought* (Princeton, NJ: Princeton University Press, 2003); Carl Schorske, *Fin-de-Siècle Vienna: Politics and Culture* (New York: Vintage Books, 1981); Marsha Rozenblit, *The Jews of Vienna, 1867–1914: Assimilation and Identity* (Albany: State University of New York Press, 1984); Marsha Rozenblit, *Reconstructing a National Identity: The Jews of Habsburg Austria during World War I* (Oxford: Oxford University Press, 2001).

58. Ash, *Gestalt Psychology in German Culture, 1890–1967*, 28–41.

59. Mitchell Ash explains that this collection included a set of tuning forks donated by Joseph Joachim, a "pipe organ" of glass tubing designed by Stumpf, and a "tone variator" invented by William Stern to electronically regulate the presentation of ascending and descending series of tones (see Ash, *Gestalt Psychology in German Culture, 1890–1967*, 39).

60. Ash, *Gestalt Psychology in German Culture, 1890–1967*, 28–41.

61. Mitchell Ash argues that both the Berlin and Leipzig laboratories employed careful introspection with the goal of precise characterization of the phenomena of investigation and an "elite of suitable experimenters" to be the "natural-scientific version of the German universities' primary purpose, the training of elite civil servants" (Ash, *Gestalt Psychology in German Culture, 1890–1967*, 40).

62. By the outbreak of World War II, the Phonogramm-Archiv maintained a collection of over 13,000 cylinders. Following the evacuation of the Phonogramm-Archiv in 1944 it was long thought that the collection was lost. A diligent effort following the reunification of Germany has resulted in a restoration of nearly the entire prewar collection. The Department for Ethnomusicology at the Ethnographical Museum in Berlin now maintains the Phonogramm-Archiv.

63. Otto Abraham and Erich Moritz von Hornbostel, "Über die Bedeutung des Phonographen für vergleichende Musikwissenschaft," *Zeitschrift für Ethnologie* 36 (1904): 226–227.

64. See Bartók's discussion of the advantages of the phonograph and gramophone in ethnomusicological studies in "Preface to *Romanian Folk Songs from Bihor County*," in *Béla Bartók: Studies in Ethnomusicology*, ed. Benjamin Suchoff, 1–23 (Lincoln: University of Nebraska Press, [1913] 1997), as well as "Why and How Do We Collect Folk Music?" and "Some Problems of Folk Music Research in East Europe," in *Béla Bartók Essays* , ed. Benjamin Suchoff, 9–24, 173–194 (Lincoln: University of Nebraska Press, 1992).

65. Bartók, "Why and How Do We Collect Folk Music?", 14.

66. Bartók, "Some Problems of Folk Music Research in East Europe," 175.

67. Bartók, "Why and How Do We Collect Folk Music?" 14.

68. The phonograph could not, however, be used exclusively. Bartók had found that a singer, when asked to sing before the phonograph, was likely to alter his melody "in spite of himself." For this very reason as well as the elasticity of the folk-song form, Bartók continued, "the researcher must locate and control these alterations." He also noted that the singer would perform his song more slowly and with more ornamentation for the phonograph, often with the consequence of generating more errors due to his "momentary loss of memory" (see "Preface to *Romanian Folk Songs from Bihor County*," in *Béla Bartók: Studies in Ethnomusicology*, ed. Benjamin Suchoff, 1–24 (Lincoln: University of Nebraska Press, [1913] 1997), 1).

69. Bartók, "Why and How Do We Collect Folk Music?", 10.

70. Stumpf, *Die Anfänge der Musik*, 7–8.

71. Darwin's theory of the origin of human music was rooted in his theory of sexual selection. Music originated with courtship and mating rituals. This contrasted with Herbert Spencer's assertion that music developed from spoken communication. Perhaps most influential in Germany was Karl Bücher's approach, locating the rhythms of music in the repetitive motions of human labor. While each of these theories attached the origin of music to universal human principles of reproduction, language, or repetitive labor, they did little to explain the origin of musical qualities, to explain how music evolved the way it did (see Stumpf, *Die Anfänge der Musik*, 8–22, as well as Alexander Rehding, "The Quest for the Origins of Music in Germany circa 1900," *Journal of the American Musicological Society* 53, no. 2 (2000): 345–385).

72. Stumpf, *Die Anfänge der Musik*, 31, 35, 85–86.

73. Stumpf, *Die Anfänge der Musik*, 85.

74. Stumpf, *Die Anfänge der Musik*, 85.

75. Stumpf, *Die Anfänge der Musik*, 86.

76. Steege, "*Bel Canto* Refracted," 24–25; Timothy Armstrong, *Modernism, Technology, and the Body: A Cultural Study* (Cambridge: Cambridge University Press, 1998).

77. Steege, "*Bel Canto* Refracted," 25.

Coda

1. Thomas Mann, *Doctor Faustus: The Life of the German Composer Adrian Leverkühn as Told by a Friend* (New York: Knopf, [1948] 1999), 534.

2. Rose Subotnik, "Adorno's Diagnosis of Beethoven's Late Style," *Journal of the American Musicological Society* 29, no. 2 (1976): 242–275 (quote on 245).

3. Subotnik refers to Adorno's "Spätstil" and "Verfremdetes Hauptwerk" (Subotnik, "Adorno's Diagnosis of Beethoven's Late Style," 261).

4. Subotnik, "Adorno's Diagnosis of Beethoven's Late Style," 244.

5. Subotnik, "Adorno's Diagnosis of Beethoven's Late Style," 265.

6. Edward Said, "Adorno as Lateness Itself," in Nigel Gibson and Andrew Rubin, eds., *Adorno: A Critical Reader*, 193–208 (London: Blackwell, 2002).

7. Said, "Adorno as Lateness Itself," 207–208.

8. Subotnik, "Adorno's Diagnosis of Beethoven's Late Style," 245.

9. Theodor Adorno, *Philosophy of Modern Music* (New York: Continuum, [1949] 2004), 133.

10. Burkhard Bilger, "Profiles: The Possibilian," *New Yorker*, April 25,2011, 54–65.

11. International Network for Neuroaesthetics, http://neuroaesthetics.net.

12. Oliver Sacks, *Musicophilia: Tales of Music and the Brain* (New York: Knopf, 2007), 212.

13. Alex Ross, *The Rest Is Noise: Listening to the Twentieth Century* (New York: Farrar, Straus and Giroux, 2007), xviii.

References

Abraham, Otto, and Erich Moritz von Hornbostel. 1904. Über die Bedeutung des Phonographen für vergleichende Musikwissenschaft. *Zeitschrift fur Ethnologie* 36: 222–236.

Adorno, Theodor. [1949] 2004. *Philosophy of Modern Music.* New York: Continuum.

Applegate, Celia. 2005. *Bach in Berlin: Nation and Culture in Mendelssohn's Revival of the St. Matthew Passion.* Ithaca, NY: Cornell University Press.

Applegate, Celia, and Pamela Potter. 2002. Germans as the "people of music": Genealogy of an identity. In Celia Applegate and Pamela Potter, eds., *Music and German National Identity.* 1–35. Chicago: University of Chicago Press.

Applegate, Celia, and Pamela Potter, eds. 2002. *Music and German National Identity.* Chicago: University of Chicago Press.

Arendt, Hans-Jürgen. 2003. Gustav Theodor Fechner und Hermann Härtel—eine lebenslange Freundschaft. In Ulla Fix, Irene Altmann, and Gustav Fechner, eds., *Fechner und die Folgen außerhalb der Naturwissenschaften: Interdisziplinäres Kolloquium zum 200. Geburtstag Gustav Theodor Fechners,* 233–244. Tübingen: Max Niemeyer Verlag.

Armstrong, Timothy. 1998. *Modernism, Technology, and the Body: A Cultural Study.* Cambridge: Cambridge University Press.

Arnzt, Michael. 2001. "Nehmen Sie Riemann ernst?" Zur Bedeutung Hugo Riemanns für die Emanzipation der Musik. In Tatjana Böhme-Mehner and Klaus Mehner, eds., *Hugo Riemann (1849–1919): Musikwissenschaftler mit Universalanspruch,* 9–16. Vienna: Bölhau.

Ash, Mitchell. 1995. *Gestalt Psychology in German Culture, 1870–1967: Holism and the Quest for Objectivity.* Cambridge: Cambridge University Press.

Bartók, Béla. [1913] 1997. Preface to *Romanian Folk Songs from Bihor County.* In Benjamin Suchoff, ed., *Béla Bartók: Studies in Ethnomusicology,* 1–23. Lincoln: University of Nebraska Press.

Bartók, Béla. 1981. *The Hungarian Folk Song*. Benjamin Suchoff, ed., M. D. Calvocoressi, trans. Albany: State University of New York Press.

Bartók, Béla. [1992] 1997. Why and how do we collect folk music? and Some problems of folk music research in East Europe. Reprinted in Benjamin Suchoff, ed., *Béla Bartók Essays*. 9–24, 173–194. Lincoln: University of Nebraska Press.

Bijsterveld, Karin. 2001. The diabolical symphony of the mechanical age: Technology and symbolism of sound in European and North American noise abatement campaigns, 1900–40. *Social Studies of Science* 31:37–70.

Bijsterveld, Karin. 2006. Listening to machines: Industrial noise, hearing loss and the cultural meaning of sound. *Interdisciplinary Science Reviews* 31:323–377.

Bilger, Burkhard. 2011. Profiles: The possibilian. *New Yorker,* April 25, 54–65.

Blackmore, John. 1972. *Ernst Mach: His Work, Life, and Influence*. Berkeley: University of California Press.

Blasius, Leslie. 2001. Nietzsche, Riemann, Wagner: When music lies. In Suzannah Clark and Alexander Rehding, eds., *Music Theory and Natural Order from the Renaissance to the Early Twentieth Century*, 93–107. Cambridge: Cambridge University Press.

Bohlman, Philip. 2000a. Composing the cantorate: Westernizing Europe's Other within. In Georgina Born and David Hesmondhalgh, eds., *Western Music and Its Others*. Berkeley: University of California Press.

Bohlman, Philip V. 2000b. The remembrance of things past: Music, race, and the end of history in Modern Europe. In Ronald Radano and Philip V. Bohlman, eds., *Music and the Racial Imagination,* 644–676. Chicago: University of Chicago Press.

Borchard, Beatrix. 2012. Joseph Joachim. *Grove Music Online.* http://www.oxfordmusiconline.com.

Borchmeyer, Dieter. 2003. *Drama and the World of Richard Wagner*. Princeton, NJ: Princeton University Press.

Boring, Edwin. 1929. The psychology of controversy. *Psychological Review* 36: 97–121.

Botstein, Leon. 1990. Time and memory in Brahms's Vienna. In Walter Frisch, ed., *Brahms and His World,* 3–22. Princeton, NJ: Princeton University Press.

Bowler, Peter. [1989] 2003. *Evolution: The History of an Idea*. 3rd ed. Berkeley: University of California Press.

Brahms, Johannes, and Joseph Joachim. [1860] 1990. Appendix 2: Manifesto. David Brodbeck. "Brahms, the Third Symphony, and the New German School." In Walter Frisch, ed., *Brahms and His World,* 79–80. Princeton, NJ: Princeton University Press. [Original: *Berliner Musik-Zeitung Echo,* May 6, 1860.]

Brain, Robert. 1996. *The Graphic Method: Inscription, Visualization, and Measurement in Nineteenth-Century Science and Culture.* Doctoral dissertation, University of California at Los Angeles.

Brain, Robert. 1998. Standards and semiotics. In Timothy Lenoir, ed., *Inscribing Science: Scientific Texts and the Materiality of Communication*, 249–284. Stanford, CA: Stanford University Press.

Brock, Adrian, ed. 2006. *Internationalizing the History of Psychology.* New York: New York University Press.

Brodbeck, David. 1990. Brahms, The Third Symphony, and the New German School. In Walter Frisch, ed., *Brahms and His World.* Princeton, NJ: Princeton University Press.

Brown, Julie. 2000. Bartók, the Gypsies, and hybridity in music. In Georgina Born and David Hesmondhalgh, eds., *Western Music and Its Others: Difference, Representation, and Appropriation in Music*, 119–142. Berkeley: University of California Press.

Burnham, Scott. 1989. The role of sonata form in A. B. Marx's theory of form. *Journal of Music Theory* 33 (2): 247–271.

Burnham, Scott. 1992. Method and motivation in Hugo Riemann's history of music theory. *Music Theory Spectrum* 14:4–9.

Burnham, Scott. 1993. Musical and intellectual values: Interpreting the history of tonal theory. *Current Musicology* 53:76–88.

Burnham, Scott, ed. and trans. 2006. *Musical Form in the Age of Beethoven: Selected Writings on Theory and Method.* Cambridge: Cambridge University Press.

Buss, Allan R., ed. 1979. *Psychology in Social Context.* New York: Irvington.

Cahan, David. 2010. Helmholtz in Gilded-Age America: The International Electrical Congress of 1893 and the relations of science and technology. *Annals of Science* 67:1–38.

Cambrosio, Alberto, Daniel Jacobi, and Peter Keating. 1993. Ehrlich's "Beautiful Pictures" and the controversial beginnings of immunological imagery. *Isis* 84: 662–699.

Chladni, E. F. F. 1802. *Die Akustik.* Leipzig: Breitkopf und Härtel.

Chladni, E. F. F. 1826. Wellenlehre, auf Experimente gegründet, oder über die Wellen tropfbarer Flüssigkeiten, mit Anwendung auf die Schall- und Lichtwellen. *Cäcilia: Eine Zeitschrift für die musikalische Welt* 4: 189–212.

Coen, Deborah. 2007. *Vienna in the Age of Uncertainty: Science, Liberalism, and Private Life.* Chicago: University of Chicago Press.

Collins, Harry. 1992. *Changing Order: Replication and Induction in Scientific Practice*. Chicago: University of Chicago Press.

Cook, Nicholas. 2007. *The Schenker Project: Culture, Race and Music Theory in Fin-de-Siècle Vienna*. Oxford: Oxford University Press.

Crary, Jonathan. 1999. *Suspensions of Perception: Attention, Spectacle, and Modern Culture*. Cambridge, MA: MIT Press.

Dahlhaus, Carl. 1957. War Zarlino Dualist? *Die Musikforschung* 10:286–290.

Dahlhaus, Carl. [1982] 1995. *Esthetics of Music*. Trans. William W. Austin. Cambridge: Cambridge University Press.

Dahlhaus, Carl. 1989. *Nineteenth-Century Music*. Trans. J. Bradford Robinson. Berkeley: University of California Press.

Danziger, Kurt. 1990. *Constructing the Subject: Historical Origins of Psychological Research*. Cambridge: Cambridge University Press.

Daston, Lorraine, and Peter Galison. 2007. *Objectivity*. New York: Zone Books.

Debussy, Claude. 1977. *Debussy on Music*. Trans. Richard Langham Smith. New York: Knopf.

Duffin, Ross. 2007. *How Equal Temperament Ruined Harmony (and Why You Should Care)*. New York: Norton.

Ellis, Alexander. [1877] 1954. Appendix XI: Vibration of the membrana basilaris in the cochlea. In Hermann Helmholtz, *On the Sensations of Tone as a Physiological Basis for the Theory of Music*, 2nd English ed., translated, revised, and corrected from 4th German ed., with additional notes and appendix, by Alexander Ellis. New York: Dover.

Ellis, Alexander. 1880. On the history of musical pitch. *Journal of the Society of Arts* 28:293–336.

Ellis, Alexander. 1885. On the musical scales of various nations. *Journal of the Society of Arts* 33:485–527.

Fauser, Annegret. 2005. *Musical Encounters at the 1889 Paris World's Fair*. Rochester, NY: University of Rochester Press.

Fechner, Gustav. 1836. *Das Büchlein vom Leben nach dem Tode*. Dresden: Grimmer. [This 1st edition appeared under the name of "Dr. Mises."]

Fechner, Gustav. 1851. *Zend-Avesta oder über die Dinge des Himmels und des Jenseits. Vom Standpunkt der Naturbetrachtung*. 3 Theile. Leipzig: Leopold Voß.

Fechner, Gustav. [1860] 1864. *Elemente der Psychophysik*. Vol. 1. Amsterdam: E. J. Bonset.

Fechner, Gustav. 1871. Zur experimentellen Aesthetik, th. I. *Gesellschaft der Wissenschaften zu Leipzig* 9:553–635.

Fechner, Gustav. 1872. Letter to Helmholtz, March 12. Universitätsbibliothek Leipzig, Fechner Nachlass 42:2.

Fechner, Gustav. 1875. *Kleine Schriften*. Leipzig: Breitkopf und Härtel. [Appeared under the name Dr. Mises.]

Fechner, Gustav. [1876] 1925. *Vorschule der Aesthetik*. 3rd ed. Leipzig: Breitkopf und Härtel.

Fechner, Gustav. [1879] 1994. Die Tagesansicht gegenüber die Nachtansicht. Eschborn: Klotz. Original: Leipzig: Breitkopf und Härtel.

Fechner, Gustav. 1889. *Elemente der Psychophysik*. Vol. 2. Leipzig: Breitkopf und Härtel.

Fechner, Fechner. 1966. Elements of Psychophysics. Vol. 1 Trans. Helmut Adler, ed. Davis Howes and Edwin Boring. New York: Holt, Rinehart and Winston.

Fechner, Gustav. 2004. *Tagebücher 1828 bis 1879*. Vols. 1–2. Leipzig: Verlag der Sächsischen Akademie der Wissenschaften.

Fix, Ulla, and Irene Altmann, eds. 2003. *Fechner und die Folgen außerhalb der Naturwissenschaften: Interdisziplinäres Killowuium zum 200. Geburstag Gustav Theodor Fechners*. Tübingen: Max Niemeyer Verlag.

Fostle, D. W. 1995. *The Steinway Saga: An American Dynasty*. New York: Scribner.

Galison, Peter. 1997. *Image and Logic: A Material Culture of Microphysics*. Chicago: University of Chicago Press.

Garratt, James. 2010. *Music, Culture and Social Reform in the Age of Wagner*. Cambridge: Cambridge University Press.

Grey, Thomas S. 2012. Eduard Hanslick. *Grove Music Online*. http://www.oxfordmusiconline.com.

Hagner, Michael. 2003. Toward a history of attention in culture and science. *MLN* 118 (3): 670–687.

Hall, G. Stanley. 1912. *Founders of Modern Psychology*. New York: Appleton.

Hällstrom, Gustav. 1832. Von den Combinationstönen. Annalen, 2nd series, 24:438–466.

Hanslick, Eduard. [1854a] 1918. *Vom Musikalisch-Schönen: Ein Beitrag zur Revision der Ästhetik der Tonkunst*. 12th ed. Leipzig: Breitkopf und Härtel.

Hanslick, Eduard. [1854b] 1957. *The Beautiful in Music*. Ed. Morris Weitz, trans. Gustav Cohen from the 7th German ed. in 1891. New York: Liberal Arts Press.

Hanslick, Eduard. [1861] 1950. Joseph Joachim. In Henry Pleasants, ed. and trans., Music Criticisms, 1846–1899, 78–81. New York: Penguin.

Hanslick, Eduard. 1884. *Ein Brief über die "Clavierseuche." Suite. Aufsätze über Musik und Musiker.* Techen: K. Prochaska.

Hanslick, Eduard. [1862] 1950. Brahms. In Henry Pleasants, ed. and trans., Music Criticisms, 1846–1899, 82–86. New York: Penguin.

Hanslick, Eduard. 1899. Robert Schumann in Endenich. In Eduard Hanslick, *Am Ende des Jahrhunderts, 1895–1899*, 317–342. 2nd ed. Berlin: Allgemeiner Verein für Deutsche Litteratur.

Hatfied, Gary. 1990. *The Natural and the Normative: Theories of Spatial Perception from Kant to Helmholtz*. Cambridge, MA: MIT Press.

Hatfield, Gary. 1993. Helmholtz and classicism: The science of aesthetics and the aesthetics of science. In David Cahan, ed., *Hermann von Helmholtz and the Foundations of Nineteenth-Century Science*, 522–558. Berkeley: University of California Press.

Hauptmann, Moritz. 1853. Die Natur der Harmonik und Metrik. Leipzig: Breitkopf und Härtel.

Heidelberger, Michael. 1993a. Force, law, and experiment: The evolution of Helmholtz's philosophy of science. In David Cahan, ed., *Hermann von Helmholtz and the Foundations of Nineteenth-Century Science*, 461–497. Berkeley: University of California Press.

Heidelberger, Michael. 1993b. *Die innere Seite der Natur: Gustav Theodor Fechners wissenschaftliche-philosophische Weltauffassung*. Frankfurt am Main: Vittorio Klostermann.

Heidelberger, Michael. 2004. *Nature from Within: Gustav Theodor Fechner and His Psychophysical Worldview.* Trans. Cynthia Klohr. Pittsburgh: University of Pittsburgh Press.

Helmholtz, Ferdinand. 1837. Die Wichtigkeit der allgemeine Erziehung für das Schöne. *Zu der öffentlichen Prüfung der Zöglinge des hiesigen Königlichen Gymnasiums den 21sten und 22sten März laden ganz ergebenst ein Director und Lehrercollegium*, 1–44. Potsdam: Decker'schen Geheimen Oberhofbuchdruckerei-Establissement.

Helmholtz, Hermann. 1855. Über das Sehen des Menschen. In *Vorträge und Reden*, vol. 1, 115–116. Braunschwieg: F. Vieweg und Sohn.

Helmholtz, Hermann. 1856. Ueber Combinationstöne. *Poggendorff's Annalen der Physik und Chemie* 99:497–540.

Helmholtz, Hermann. [1857a] 1884. Ueber die physiologischen Ursachen der musikalischen Harmonie. In *Vorträge und Reden*, vol. 1, 99–155. Braunschweig: F. Vieweg und Sohn.

Helmholtz, Hermann. [1857b] 1903. On the physiological causes of harmony in music. In *Popular Lectures on Scientific Subjects*, 53–94. London: Longmans, Green, and Co.

Helmholtz, Hermann. 1860a. On the motion of the strings of a violin. *Proceedings of the Philosophical Society of Glasgow.* Glasgow: Philosophical Society, *17–21.*

Helmholtz, Hermann. 1860b. Über musikalische Temperatur. Paper presented on November 23. *Naturh.-med. Verein in Heidelberg* 2:73–75.

Helmholtz, Hermann. 1862. Über die arabisch-persische Tonleiter. Paper presented on May 31. *Naturh.-med. Verein in Heidelberg* 2:216–217.

Helmholtz, Hermann. [1863] 1870. *Die Lehre von den Tonempfindungen als physiologische Grundlage für die Theorie der Musik.* 3rd ed. Braunschweig: Druck und Verlag von Friedrich Vieweg und Sohn.

Helmholtz, Hermann. [1868] 1884. Die neueren Fortschritte in der Theorie des Sehen. *Vorträge und Reden*, vol. 1. 233–332. Wiesbaden: F. Vieweg.

Helmholtz, Hermann. [1869] 1904. The aim and progress of physical science: An opening address delivered at the Naturforscher Versammlung in Innsbrück. In Hermann Helmholtz, *Popular Lectures on Scientific Subjects*, trans. E. Atkinson. London: Longmans, Green, and Co.

Helmholtz, Hermann. [1877] 1912. *On the Sensations of Tone as a Physiological Basis for the Theory of Music.* 4th English ed., translated, revised, and corrected from 4th German ed., with additional notes and appendix, by Alexander Ellis. London: Longmans, Green, and Co.

Helmholtz, Hermann. 1990. *Letters of Hermann von Helmholtz to His Wife, 1847–1859.* Ed. Richard L. Kremer. Stuttgart: Franz Steiner Verlag.

Helmholtz, Hermann. 1993. *Letters of Hermann von Helmholtz to His Parents: The Medical Education of a German Scientist, 1837–1846.* Ed. David Cahan. Stuttgart: Franz Steiner Verlag.

Hering, Ewald. 1911. On memory as a general function of organized matter. In Samuel Butler, ed., *Unconscious Memory*, 80–83. 2nd ed. New York: Dutton.

Hiebert, Elfrieda. 2004. *Helmholtz's musical acoustics: Incentive for practical techniques in pedaling and touch at the piano. Papers Read at the IMS Intercongressional Symposium, August 2004. Budapest: Liszt Ferenc Academy of Music.*

Hiebert, Elfrieda, and Erwin Hiebert. 1994. Musical thought and practice: Links to Helmholtz's *Tonempfindungen*. In Lorenz Krüger, ed., *Universalgenie Helmholtz: Rückblick nach 100 Jahren*, 295–314. Berlin: Akademie Verlag.

Holmes, Frederic, and Kathryn Olesko. 1995. The images of precision: Helmholtz and the graphical method in physiology. In M. Norton Wise, ed., *The Values of Precision*, 198–221 Princeton, NJ: Princeton University Press.

Jackson, Myles. 2006. *Harmonious Triads: Physicists, Musicians, and Instrument Makers in Nineteenth-Century Germany*. Cambridge, MA: MIT Press.

James, William. [1904] 2005. Introduction. In Gustav Fechner, *The Little Book of Life After Death*, trans. Mary C. Wadsworth. Boston: Weiser Books.

James, William. 1909. Concerning Fechner. In William James, *A Pluralistic Universe: Hibbert Lectures at Manchester College*. London: Longmans, Green & Co.

Joachim, Joseph. 1914. *Letters from and to Joseph Joachim*. Ed. and trans. Nora Bickley. New York: Macmillan.

Johnson, James. 1995. *Listening in Paris: A Cultural History*. Berkeley: University of California Press.

Jorgensen, Owen. 1991. *Tuning: Containing the Perfection of Eighteenth-Century Temperament, the Lost Art of Nineteenth-Century Temperament, and the Science of Equal Temperament*. East Lansing: Michigan State University Press.

Jurkowitz, Edward. 2003. Helmholtz and the liberal unification of science. *Historical Studies in the Physical Sciences* 32:291–317.

Karpath, Ludwig. 1972. Letter part of Ernst Anton Lederer's (Mach's grandson's) private collection at his home in Essex Fells, New Jersey. Quoted (and translated) in John Blackmore, *Ernst Mach: His Work, Life, and Influence*. Berkeley: University of California Press.

Kelly, Alfred. 1981. *Descent of Darwin: The Popularization of Darwinism in Germany, 1860–1914*. Chapel Hill: University of North Carolina Press.

Kennaway, James. 2010. From sensibility to pathology: The origins of the idea of nervous music around 1800. Journal of the History of Medicine and Allied Sciences 65: 396–426.

Koenigsberger, Leo. 1906. *Hermann von Helmholtz*. Trans. Frances A. Welby. Oxford: Clarendon Press.

Kösser, Uta. 2003. Fechners Ästhetik im Kontext. In Ulla Fix and Irene Altmann, eds., *Fechner und die Folgen außerhalb der Naturwissenschaften: Interdisziplinäres Kolloquium zum 200. Geburstag Gustav Theodor Fechner*. Tübingen: Max Niemeyer Verlag.

Kramer, Lawrence. 2002a. Franz Liszt and the virtuoso public sphere: Sight and sound in the rise of mass entertainment. In Lawrence Kramer, *Musical Meaning: Toward a Critical History*, 68–99. Berkeley: University of California Press.

Kramer, Lawrence. 2002b. *Musical Meaning: Toward a Critical History*. Berkeley: University of California Press.

Kulke, Eduard. 1872. Letter to Ernst Mach, October 26. Ernst Mach Nachlass, Deutsches Museum Archives, Munich.

Kulke, Eduard. 1884. *Über die Umbildung der Melodie: Ein Beitrag zur Entwickelungslehre*. Prague: J. G. Calve'sche K. K. Hof- und Univ.-Buchhandlung.

Kulke, Eduard. 1906. *Kritik der Philosophie des Schönen*. Leipzig: Deutsche Verlagactiengesellschaft.

Kursell, Julia. 2006. *Sound objects. Presented at "Sounds of Science, Schall im Labor (1800–1930)" Workshop, Max-Planck-Institut für Wissenschaftsgeschichte, Berlin, October 5–7.*

Large, David C., and William Weber, eds. 1984. *Wagnerism in European Culture and Politics*. Ithaca, NY: Cornell University Press.

Latour, Bruno, and Steve Woolgar. 1986. *Laboratory Life: The Construction of Scientific Facts*. Princeton, NJ: Princeton University Press.

Lenoir, Timothy. 1993. The eye as mathematician: Clinical practice, instrumentation, and Helmholtz's construction of an empiricist theory of vision. In David Cahan, ed., *Hermann von Helmholtz and the Foundations of Nineteenth-Century Science*, 109–153. Berkeley: University of California Press.

Lenoir, Timothy. 1997a. *Instituting Science: The Cultural Production of Scientific Disciplines*. Stanford, CA: Stanford University Press.

Lenoir, Timothy. 1997b. The politics of vision: Optics, painting, and ideology in Germany, 1845–1895. In Timothy Lenoir, *Instituting Science: The Cultural Production of Scientific Disciplines*, 131–178. Stanford, CA: Stanford University Press.

Lenoir, Timothy. 1998. Inscription practices and materialities of communication. In Timothy Lenoir, ed., *Inscribing Scientific Texts and the Materiality of Communication*, 1–19. Stanford, CA: Stanford University Press.

Le Rider, Jacques. 1993. *Modernity and Crises of Identity: Culture and Society in Fin-de-Siècle Vienna*. Trans. Rosemary Morris. New York: Continuum.

Lieberman, Richard K. 1995. *Steinway and Sons*. New Haven, CT: Yale University Press.

Lindley, Mark. 1984. *Lutes, Viols, and Temperaments*. Cambridge: Cambridge University Press.

Lindley, Mark, and Ronald Turner-Smith. 1993. *Mathematical Models of Musical Scales*. Bonn: Verlag für systematische Musikwissenschaft.

Lorenz, Carl. 1890. Untersuchungen über die Auffassung von Tondistanzen. *Philosophische Studien* 6:26–103.

Lunn, Henry C. 1879. Free and cheap concerts for the people. *Musical Times* 20:306–307.

Lynch, Michael. 1985. *Art and Artifact in Laboratory Science: A Study of Shop Work and Shop Talk in a Research Laboratory*. London: Routledge & Kegan Paul.

Lynch, Michael. 1988. The externalized retina: Selection and mathematization in the visual documentation of objects in the life sciences. In Michael Lynch and Steve Woolgar., eds., *Representation in Scientific Practice,* 153–186. Cambridge, MA: MIT Press.

Mach, Ernst. 1863. Zur Theorie des Gehörorgans. Sitzungsberichte der kaiserlichen Akademie der Wissenschaften 48 (2): 283–300.

Mach, Ernst. 1864. Über einige der physiologischen Akustik angehörige Erscheinungen. Sitzungsberichte der kaiserlichen Akademie der Wissenschaften 50:342–362.

Mach, Ernst. 1865a. Bemerkungen über die Accommodation des Ohres. *Sitzungsberichte der Kaiserlichen Akademie der Wissenschaften* 51:343–346.

Mach, Ernst. 1865b. Die Erklärung der Harmonie. In Ernst Mach, *Zwei populäre Vorlesungen über musikalische Akustik.* Graz: Leuschner & Lubensky.

Mach, Ernst. 1866. *Einleitung in die Helmholtz'sche Musiktheorie: Populär für Musiker dargestellt.* Graz: Leuschner & Lubensky.

Mach, Ernst. 1871. Ueber die physikalische Bedeutung der Gesetze der Symmetrie. Lotos. *Zeitschrift für Naturwissenschaften* 21:139–147.

Mach, Ernst. 1872a. *Die Geschichte und die Wurzel des Satzes von der Erhaltung der Arbeit.* Prague: J. G. Calve'sche Univ.-Buchhandl.

Mach, Ernst. 1872b. Letter to Eduard Kulke. May 30, letter (no. 4). Ernst Mach Papers, Dibner Library of the History of Science and Technology, Smithsonian Institution Special Collections, Washington, DC.

Mach, Ernst. 1876–1878. Letter to Eduard Kulke. Undated (though likely between 1876 and 1878), letter (no. 24). Ernst Mach Papers, Dibner Library of the History of Science and Technology, Smithsonian Institution Special Collections, Washington, DC.

Mach, Ernst. 1877. Notizbuch 514. Ernst Mach Nachlass, Deutsches Museum, Munich.

Mach, Ernst. 1882. Die Ökonomische Natur der physikalischen Forschung. *Almanach der Kaiserlichen Akademie der Wissenschaft* 32: 293–319.

Mach, Ernst. [1886a] 1902. *Die Analyse der Empfindungen und das Verhältnis des Physischen zum Psychischen.* 8th ed. Jena: G. Fischer.

Mach, Ernst. [1886b] 1959. *The Analysis of the Sensations and the Relation of the Physical to the Psychical.* Trans. C. M. Williams from the 5th (1897) German ed. New York: Dover.

Mach, Ernst. 1892. Why has man two eyes? In Ernst Mach, *Popular Scientific Lectures.* Trans. Thomas McCormack, 66–88. Chicago: Open Court.

Mach, Ernst. 1896a. Uber die Cortischen Fasern des Ohres. Reprinted in Ernst Mach, *Populär-Wissenschaftliche Vorlesungen*. Leipzig: J. A. Barth.

Mach, Ernst. 1896b. Die Erklärung der Harmonie. Reprinted in Ernst Mach, *Populär-Wissenschaftliche Vorlesungen*. Leipzig: J. A. Barth.

Mach, Ernst. 1896c. Postcard to Schenker dated December 2. Held in the Oswald Jonas Memorial Collection, Special Collections, University of California, Riverside Libraries, box 12, folder 47. Available online at http://mt.ccnmtl.columbia.edu/schenker/correspondence/postcard/oj_1247_12296.html.

Mach, Ernst. 1898. On symmetry. In Ernst Mach, Popular Scientific Lectures. 3rd ed. Trans. Thomas McCormack, 89–106. Chicago: Open Court.

Mach, Ernst. 1906. Foreword. In Eduard Kulke, *Kritik der Philosophie des Schönen*. Leipzig: Deutsche Verlagactiengesellschaft.

Mach, Ernst. n.d. Letter to Eduard Kulke. Undated (likely early 1860s), letter (no. 1). Ernst Mach Papers, Dibner Library of the History of Science and Technology, Smithsonian Institution Special Collections, Washington, DC.

Mach, Ernst, and Johann Kessel. 1872a. Die Function der Trommelhöhle und der Tuba Eustachii. *Sitzungsberichte der kaiserlichen Akademie der Wissenschaften* 66:329–336.

Mach, Ernst, and Johann Kessel. 1872b. Versuche über die Accommodation des Ohres. *Sitzungsberichte der kaiserlichen Akademie der Wissenschaften* 66:337–343.

Mann, Thomas. [1918] 1974. Betrachtungen eines Unpolitischen. Vol. 12. Gesammelte Werke. Frankfurt: S. Fischer.

Mann, Thomas. [1948] 1999. *Doctor Faustus: The Life of the German Composer Adrian Leverkühn as Told by a Friend*. New York: Knopf.

Marx, Adolf. 1824. Ueber die Anforderungen unserer Zeit an musikalscihe Kritik; in besonderm Bezuge auf diese Zeitung. *Berliner allgemeine musikalische Zeitung* no. 1:2–4, no. 2:9–11, no. 3:17–19.

Marx, Adolf. 1826. Standpunkt der Zeitung. *Berliner allgemeine musikalische Zeitung* 3 (52): 421–424.

Marx, Adolf. [1841] 2006. The old school of music in conflict with our times. In Scott Burnham, ed. and trans., Musical Form in the Age of Beethoven: Selected Writings on Theory and Method, 17–34. Cambridge: Cambridge University Press. Original: *Die alte Musiklehre im Streit mit unserer Zeit*. Leipzig: Breifkopf und Härtel.

Marx, Adolf. [1839] 1852. The Universal School of Music: A Manual for Teachers and Students in Every Branch of Musical Art. Trans. from the 5th ed. of the original by A. H. Wehrman. London: Robert Cocks and Co. Original: *Allgemeine Musiklehre*. Leipzig: Breitkopf und Härtel.

Marx, Adolf. 1854. Die Musik des neunzehnten Jahrhunderts und ihre Pflege: Methode der Musik. Leipzig: Breitkopf und Härtel.

Marx, Adolf. [1856] 2006. Form in music. In Scott Burnham, ed. and trans., *Musical Form in the Age of Beethoven: Selected Writings on Theory and Method,* 55–90. Cambridge: Cambridge University Press. Original: Die Form in Musik. In J. A. Romberg, ed., *Die Wissenschaften in neunzehnten Jahrhundert,* vol. 2. Leipzig: J. A. Romberg's Verlag.

Mendel, Hermann, ed. 1870–1888. *Musikalisches Conversations-Lexikon.* Vols. 1–11 and suppl. vol. Berlin: R. Oppenheim.

Mody, Cyrus. 2005. The sounds of science: Listening to laboratory practice. *Science, Technology & Human Values* 30 (2): 175–198.

Münnich, Richard. 1909. Von Entwicklung der Riemannschen Harmonielehre und ihrem Verhältnis zu Oettingen und Stumpf. In Carl Mennicke, ed., *Riemann-Festschrift,* 60–76. Leipzig: Max Hesse.

Myers, David. 2003. *Resisting History: Historicism and Its Discontents in German-Jewish Thought.* Princeton, NJ: Princeton University Press.

Nettl, Bruno, and Philip V. Bohlman, eds. 1991. *Comparative Musicology and Anthropology of Music: Essays on the History of Ethnomusicology.* Chicago: University of Chicago Press.

Nietzsche, Friedrich. [1886] 1966. *Beyond Good and Evil: Prelude to a Philosophy of the Future.* Trans. Walter Kaufmann. New York: Vintage Books.

Nietzsche, Friedrich. [1887] 1998. *On the Genealogy of Morality: A Polemic.* Indianapolis: Hackett.

Nietzsche, Friedrich. [1888a] 1992. The Case of Wagner. In Friedrich Nietzsche, *Basic Writings of Nietzsche.* Trans. Walter Kaufmann. New York: Modern Library.

Nietzsche, Friedrich. 1888b. *Der Fall Wagner: Ein Musikanten-Problem.* Leipzig: Verlag von C. G. Neumann.

Nordau, Max. [1892] 1993. Degeneration. Reprinted from English-language ed. published in 1895; translated from 2nd ed. of the German work. Lincoln: University of Nebraska Press.

Nyhart, Lynn. 1995. *Biology Takes Form: Animal Morphology and the German Universities, 1800–1900.* Chicago: University of Chicago Press.

Oettingen, Arthur von. 1866. *Harmoniesystem in dualer Entwickelung: Studien zur Theorie der Musik.* Dorpat: W. Gläser.

Oettingen, Arthur von 1877. Wind-Componenten-Integrator. Repertorium für Meteorologie 5, no. 10.: 5007

Pantalony, David. 2005. Rudolph Koenig's workshop of sound: Instruments, theories, and the debate over combination tones. *Annals of Science* 62:57–82.

Pantalony, David. 2009. *Altered Sensations: Rudolph Koenig's Acoustical Workshop in Nineteenth-Century Paris*. Heidelberg: Springer.

Paul, Oscar. 1868. *Geschichte des Claviers vom Ursprunge bis zu den modernsten Formen dieses Instruments nebst einer Übersicht über Musikalische Abtheilung der Pariser Weltausstellung im Jahre 1867*. Leipzig: A. H. Payne.

Pearce, Trevor. 2008. Tonal functions and active synthesis: Hugo Riemann, German psychology, and Kantian epistemology. *Intégral* 22:81–116.

Pederson, Sanna. 1994 A. B. Marx, Berlin concert life, and German national identity. *19th-Century Music* 18:87–107.

Pesic, Peter. 2012. Helmholtz, Riemann, and the sirens: Sound, color, and the problem of space. Unpublished manuscript. St. John's Collage.

Pinch, Trevor, and Frank Trocco. 2002. *Analog Days: The Invention and Impact of the Moog Synthesizer*. Cambridge, MA: Harvard University Press.

Rehding, Alexander. 2000. The quest for the origins of music in Germany circa 1900. *Journal of the American Musicological Society* 53 (2): 345–385.

Rehding, Alexander. 2003. *Hugo Riemann and the Birth of Modern Musical Thought*. Cambridge: Cambridge University Press.

Review of Mach's *Einleitung in die Helmholtz'sche Musiktheorie*. 1867. *Leipziger Allgemeine Musikalische Zeitung* 7:58.

Riemann, Bernhard. 1866. Mechanik des Ohres. *Zeitschrift für rationelle Medicin*, 3rd series, 29:128–143.

Riemann, Bernhard. [1866] 1984. The mechanism of the ear. *Fusion* 6 (3): 31–38.

Riemann, Hugo. 1873. *Über das musikalische Hören*. Doctoral dissertation, Göttingen University. [Published as *Musikalische Logik: Hauptzüge der physiologischen und psychologischen Begründung unseres Musiksystems* (Leipzig: C. F. Kahnt, 1874)].

Riemann, Hugo. [1874a] 1977. Musical logic. In William Mickelsen, ed. and trans., Hugo Riemann's Theory of Harmony and History of Music Book III. Lincoln: University of Nebraska Press.

Riemann, Hugo. 1874b. *Musikalische Logik: Hauptzüge der physiologischen und psychologischen Begründung unseres Musiksystems*. Leipzig: C. F. Kahnt.

Riemann, Hugo. 1875. Die objective Existenz der Untertöne in der Schallwelle. In *Allgemeine deutsche Musikzeitung* 2: 205–206, 213–215.

Riemann, Hugo. 1877. *Musikalische Syntaxis: Grundriss einer harmonischen Satzbildungslehre*. Leipzig: Breitkopf und Härtel.

Riemann, Hugo. [1878] 1900. Der gegenwärtige Stand der musikalischen Aesthetik. In Hugo Riemann, Präludien und Studien: Gesammelte Aufsätze zur Aesthetik, Theorie, und Geschichte der Musik. Vol. 2. Leipzig: Hermann Seemann, Nachfolger.

Riemann, Hugo. 1882a. Die Natur der Harmonik. Waldersees Sammlung musikalischer Vorträge 4:159–190.

Riemann, Hugo. [1882b] 1977. The nature of harmony. In William Mickelsen, ed. and trans., Hugo Riemann's Theory of Harmony and History of Music Book III. Lincoln: University of Nebraska Press.

Riemann, Hugo. 1888. *Wie hören wir Musik?: Grundlinien der Musikästhetik*. Leipzig: Max Hesse.

Riemann, Hugo. [1889] 1901. Einige seltsame Noten bei Brahms und anderen. In Hugo Riemann, Präludien und Studien: Gesammelte Aufsätze zur Aesthetik, Theorie und Geschichte der Musik. Vol. 3. Leipzig: Hermann Seemann. Original: 1889, Musikalisches Wochenblatt.

Riemann, Hugo. 1891. *Katechismus der Akustik: Musikwissenschaft*. Leipzig: Max Hesse.

Riemann, Hugo. 1895. *Catechism of Musical Aesthetics*. London: Augener and Co. [This was the English translation of Wie hören wir Musik? Grundlinien der Musik-Ästhetik (Leipzig: Max Hesse, 1888) which also appeared under the title Katechismus der Musik-Ästhetik (Leipzig: Max Hesse, 1890)]

Riemann, Hugo. [1898] 1961. *Geschichte der Musiktheorie im IX.–XIX. Jahrhundert*. 3rd ed. Berlin: Georg Olms Verlagsbuchhandlung Hildesheim.

Riemann, Hugo. 1902a. *Grosse Kompositionslehre*. Vol. 1. Stuttgart: W. Spemann.

Riemann, Hugo, ed. 1902b. *Sechs originale chinesische und japanische Melodien*. Leipzig: Breitkopf und Härtel.

Riemann, Hugo. 1905. Das Problem des harmonischen Dualismus. *Neue Zeitschrift für Musik* 1:3.

Riemann, Hugo. 1908a. *Encyclopedic Dictionary of Music*. Trans. J. S. Shedlock. Philadelphia: Theo. Presser.

Riemann, Hugo. 1908b. *Grundriss der Musikwissenschaft*. Leipzig: Quelle & Meyer.

Riemann, Hugo. 1929. Untertöne. In Alfred Einstein *Hugo Riemanns Musiklexikon*. 11th ed. Leipzig: Max Hesse Verlag.

Ross, Alex. 2007. *The Rest Is Noise: Listening to the Twentieth Century*. New York: Farrar, Straus and Giroux.

Rozenblit, Marsha. 1984. *The Jews of Vienna, 1867–1914: Assimilation and Identity*. Albany: State University of New York Press.

Rozenblit, Marsha. 2001. *Reconstructing a National Identity: The Jews of Habsburg Austria during World War I*. Oxford: Oxford University Press.

Rudwick, Martin. 1976. The emergence of a visual language for a geological science, 1760–1840. *History of Science* 14:149–195.

Sacks, Oliver. 2007. *Musicophilia: Tales of Music and the Brain*. New York: Knopf.

Said, Edward. 1978. *Orientalism*. New York: Pantheon Books.

Said, Edward. 2002. Adorno as lateness itself. In Nigel Gibson and Andrew Rubin, eds., *Adorno: A Critical Reader*, 193–208. London: Blackwell.

Schafhäutl, Karl von. 1878. Moll und Dur in der Natur, und in der Geschichte der neuern und neuesten Harmonielehre. *Allgemeine Musikalische Zeitung* 13 (6): col. 90.

Schafhäutl, Karl von. Abhandl. *d. München. Akad.* 7:501.

Schenker, Heinrich. 1894. Das Hören in der Musik. *Neue Revue* 5:115–121.

Schenker, Heinrich. 1895. Der Geist der musikalischen Technik. *Musikalisches Wochenblatt* 26:245–246, 257–259, 273–274, 279–280, 309–310, 325–326. Reprinted in 1895 as *Der Geist der musikalischen Technik*. Leipzig: E. W. Fritsch.

Schenker, Heinrich. [1906] 1954. Harmony. Ed. Oswald Jonas, trans. Elisabeth Borgese. Chicago: University of Chicago Press. Original: Harmonielehre. Stuttgart: Cotta.

Schoenberg, Arnold. [1941] 1985. Composition with twelve tones (I). In Leonard Stein, ed., and Leo Black, trans., Style and Idea: Selected Writings of Arnold Schoenberg. Berkeley: University of California Press.

Schopenhauer, Arthur. [1818] 1969. In E. F. J. Payne, trans., *The World as Will and Representation*. Vol. 1. New York: Dover.

Schorske, Carl. 1981. *Fin-de-Siècle Vienna: Politics and Culture*. New York: Vintage Books.

Schumann, Clara. 1973. *Letters of Clara Schumann and Johannes Brahms, 1853–1896*. Vol. 2. Ed. Berthold Litzmann. New York: Vienna House.

Schumann, Robert. 1853. Neue Bahnen. *Neue Zeitschrift für Musik* 39(18): 28.

Seth, Suman. 2010. *Crafting the Quantum: Arnold Sommerfeld and the Practice of Theory, 1890–1926*. Cambridge, MA: MIT Press.

Shapin, Steve, and Simon Schaffer. 1985. *Leviathan and the Air-Pump: Hobbes, Boyle, and the Experimental Life*. Princeton, NJ: Princeton University Press.

Shaw, George Bernard. 1932. *Music in London, 1890–1894*. Vol. 2. London: Constable.

Shaw, George Bernard. 1937. *London Music in 1888–1889*. New York: Dodd, Mead, and Co.

Smith, Pamela. 2004. *The Body of the Artisan: Art and Experience in the Scientific Revolution*. Chicago: University of Chicago Press.

Smith, Pamela. 2006. Art, science, and visual culture in Early Modern Europe. *Isis* 97:83–100.

Stecker, Carl. 1890. Kritische Beiträge zu einigen Streitfragen in der Musikwissenschaft. Vierteljahrsschrift für Musikwissenschaft 6:437–465.

Steege, Benjamin. 2012. *Bel Canto* refracted: Victorian voices in the mirror. Unpublished manuscript. Columbia University.

Stumpf, Carl. 1883–1890. *Tonpsychologie*. Vols. 1–2. Leipzig: Verlag von Hirzel.

Stumpf, Carl. 1886. Alexander J. Ellis, on the musical scales of various nations. *Vierteljahrsschrift für Musikwissenschaft* 2:511–524.

Stumpf, Carl. 1888. W. Wundt, Grundzüge der physiologischen Psychologie; E. Luft, Über die Unterschiedsempfindlichkeit für Tonhöhern. *Vierteljahrsschrift für Musikwissenschaft* 4:550–559.

Stumpf, Carl. 1890. Über Vergleichungen von Tondistanzen. *Zeitschrift für Psychologie und Physiologie der Sinnesorgane* 1:431–436.

Stumpf, Carl. 1891. Wundt's Antikritik. *Zeitschrift für Psychologie und Physiologie der Sinnesorgane* 2:266–293.

Stumpf, Carl. 1892. Phonographirte Indianermelodien. *Vierteljahrsschrift für Musikwissenschaft* 8:127–144.

Stumpf, Carl. 1901. Tonsystem und Musik der Siamesen. *Beiträge zur Akustik und Musikwissenschaft* 3:69–146.

Stumpf, Carl. 1907. Letter to Joseph Joachim, dated March 5. Held in the Handschriftenarchiv und Photokopiensammlung of the Bibliotek des Staatlichen Instituts für Musikforschung, Berlin.

Stumpf, Carl. 1911. *Die Anfänge der Musik*. Leipzig: J. A. Barth.

Stumpf, Carl. [1924] 1930. *Autobiography of Carl Stumpf. History of Psychology in Autobiography*. Vol. 1. Ed. Carl Murchison, 389–441. Worcester, MA: Clark University Press. [Original: *Philosophie der Gegenwart in Selbstdarstellungen*. Vol. 5. Ed. Raymund Schmidt. Leipzig: Meiner.]

Subotnik, Rose. 1976. Adorno's diagnosis of Beethoven's late style. *Journal of the American Musicological Society* 29 (2): 242–275.

Subotnik, Rose. 1991. *Developing Variations: Style and Ideology in Western Music*. Minneapolis: University of Minnesota Press.

Tankler, Hain. 2003. Tartu and the wider world: Through international contacts to the peaks of science. *Revue de la Maison Française d'Oxford* 1 (2): 69–94.

Thompson, Emily. 1997. Dead rooms and live wires: Harvard, Hollywood, and the deconstruction of architectural acoustics, 1900–1930. *Isis* 88:597–626.

Thompson, Emily. 1999. Listening to/for modernity: Architectural acoustics and the development of modern spaces in America. In Peter Galison and Emily Thompson, eds., 352–380. *The Architecture of Science*. Cambridge, MA: MIT Press.

Thompson, Emily. 2002. *The Soundscape of Modernity: Architectural Acoustics and the Culture of Listening in America, 1900–1933*. Cambridge, MA: MIT Press.

Tremmel, Erich. 2011. Carl Emil von Schafhäutl. *Grove Music Online*. www.oxford musiconline.com.

Turner, R. Steven. 1977. The Ohm-Seebeck dispute, Hermann von Helmholtz, and the origins of physiological acoustics. *British Journal for the History of Science* 10 (34):1–24.

Turner, R. Steven. 1993. Consensus and controversy: Helmholtz on the visual perception of space. In David Cahan, ed., *Hermann von Helmholtz and the Foundations of Nineteenth-Century Science*, 154–204. Berkeley: University of California Press.

Turner, R. Steven. 1994. *In the Eye's Mind: Vision and the Helmholtz-Hering Controversy*. Princeton, NJ: Princeton University Press.

Turner, R. Steven. 1997. Demonstrating harmony: Some of the many devices used to produce Lissajous curves before the oscilloscope. *Rittenhaus* 11 (2): 33–51.

Voskul, Adelheid. 2004. Humans, machines, and conversations: An ethnographic study in the making of automatic speech recognition technologies. *Social Studies of Science* 34:393–421.

Wallaschek, Richard. 1903. *Anfänge der Tonkunst*. Leipzig: J. A. Barth.

Weber, Ernst Heinrich and Wilhelm Eduard Weber. 1825. *Die Wellenlehre auf Experimente gegründet oder über die Wellen tropfbarer Flüssigkeiten mit Anwendung auf die Schall- und Lichtwellen*. Leipzig: Gerhard Fleischer.

Weber, E. H. 1834. "De Subilitate Tactu." In E. H. Weber, *De Pulsu, Resorptione, Auditu, et Tactu*. Leipzig: Prostat Apud C. F. Koehler.

Weber, E. H. 1846. "Der Tastsinn und Gemeingefühl." In R. Wagner, ed., *Handwörterbuch der Physiologie*. Vol. 3. Abt. 2, 481–588. Braunschweig: Vieweg.

Weber, E. H. 1996. *E. H. Weber on the Tactile Senses*. Eds. Helen E. Ross and David J. Murray. East Sussex, UK: Psychology Press.

Weber, William. 1975. *Music and the Middle Class: The Social Structure of Concert Life in London, Paris, and Vienna*. New York: Holmes and Meier.

Webern, Anton. [1960] 1963. *The Path to the New Music*. Ed. Willi Reich. Bryn Mawr, PA: Theodore Presser.

Weitzmann, Carl Friedrich. [1860] 1990. Appendix 2: Parody of Brahms and Joachim's "Manifesto." In Walter Frisch, ed., *Brahms and His World*. Princeton, NJ: Princeton University Press.

Werner, Petra, and Angelika Irmscher. 1993. *Kunst und Liebe müssen sein: Briefe von Anna von Helmholtz an Cosima Wagner 1889 bis 1899*. Bayreuth: Druckhaus Bayreuth.

Wise, M. Norton. 1987. How do sums count? On the Cultural Origins of Statistical Causality. *The Probabilistic Revolution*. Vol. 1. Ed. Lorenz Krüger, Lorraine Daston, and Michael Heidelberger, 395–425. Cambridge, MA: MIT Press.

Wise, M. Norton. 2006. Making visible. *Isis* 97:75–82.

Wise, M. Norton. 2012. Bourgeois Berlin and Laboratory Science. Unpublished manuscript. University of California, Los Angeles.

Woodward, William. 1982. Wundt's Program for the New Psychology: Vicissitudes of experiment, theory, and system. In William Woodward and Mitchell G. Ash, eds., *The Problematic Science: Psychology in Nineteenth-Century Thought*, 167–197. New York: Praeger.

Woodward, William, and Mitchell G. Ash, eds. 1982. *The Problematic Science: Psychology in Nineteenth-Century Thought*. New York: Praeger.

Wundt, Wilhelm. 1874. *Grundzüge der Physiologischen Psychologie*. Leipzig: E. Engelman.

Wundt, Wilhelm. 1891. Ueber Vergleichungen von Tondistanzen. *Philosophische Studien* 6:605–640.

Wundt, Wilhelm. 1901. *Gustav Theodor Fechner: Rede zur Feier seines hundertjährigen Geburtstages*. Leipzig: Wilhelm Engelmann.

Ziemer, Hansjakob. 2008. *Die Moderne hören: Das Konzert als urbanes Forum 1890–1940*. Frankfurt: Campus.

Zimmerman, Robert. 1865. *Allgemeine Ästhetik als Formwissenschaft*. Vienna: W. Baumüller.

Zur Theorie der Musik: Die Physiker und die Musiker. 1867. *Leipziger Allgemeine Musikalische Zeitung* 21:165–169.

Index